The System of the World

Isaac Newton

ISBN-13:
978-1720489245

ISBN-10:
1720489246

The System of the World

It was the ancient opinion of not a few, in the earliest ages of philosophy, that the fixed stars stood immoveable in the highest parts of the world; that, under the fixed stars the planets were carried about the sun; that the earth, us one of the planets, described an annual course about the sun, while by a diurnal motion it was in the mean time revolved about its own axis; and that the sun, as the common fire which served to warm the whole, was fixed in the centre of the universe.

This was the philosophy taught of old by *Philolaus, Aristarchus of Samos, Plato* in his riper years, and the whole sect of the Pythagoreans; and this was the judgment of *Anaximander*, more ancient than any of them; and of that wise king of the Romans, *Numa Pompilius*, who, as a symbol of the figure of the world with the sun in the centre, erected a temple in honour of Vesta, of a round form, and ordained perpetual fire to be kept in the middle of it.

The *Egyptians* were early observers of the heavens; and from them, probably, this philosophy was spread abroad among other nations; for from them it was, and the nations about them, that the Greeks, a people of themselves more addicted to the study of philology than of nature, derived their first, as well as soundest, notions of philosophy ; and in the vestal ceremonies we may yet trace the ancient spirit of the Egyptians; for it was their way to deliver their mysteries, that is, their philosophy of things above the vulgar way of thinking, under the veil of religious rites and hieroglyphic symbols.

It is not to be denied but that *Anaxagoras, Democritus*, and others, did now and then start up, who would have it that the earth possessed the centre of the world, and that the stars of all sorts were revolved towards the west about the earth quiescent in the centre, some at a swifter, others at a slower rate.

However, it was agreed on both sides that the motions of the celestial bodies were performed in spaces altogether free and void of resistance. The whim of solid orbs was of a later date, introduced by *Eudoxus, Calippus*, and *Aristotle*; when the ancient philosophy began to decline, and to give place to the new prevailing fictions of the Greeks.

But. above all things, the phenomena of comets can by no means consist with the notion of solid orbs. The Chaldeans, the most learned astronomers of their time, looked upon the comets (which of ancient times before had been numbered among the celestial bodies) as a particular sort of planets, which, describing very eccentric orbits, presented themselves to our view only by turns, viz., once in a revolution, when they descended into the lower parts of their orbits.

And as it was the unavoidable consequence of the hypothesis of solid orbs, while it prevailed, that the comets should be thrust down below the moon, so no sooner had the late observations of astronomers restored the comets to their ancient places in the higher heavens, but these celestial spaces were at once cleared of the incumbrance of solid orbs, which by these observations were broke into pieces, and discarded for ever.

Whence it was that the planets came to be retained within any certain bounds in these free spaces, and to be drawn off from the rectilinear courses, which, left to themselves, they should have pursued, into regular revolutions in curvilinear orbits, are questions which we do not know how the ancients explained; and probably it was to give some sort of satisfaction to this difficulty that solid orbs were introduced.

The later philosophers pretend to account for it either by the action of certain vortices, as *Kepler* and *Des Cartes*; or by some other principle of impulse or attraction, as *Borelli*, *Hooke*, and others of our nation; for, from the laws of motion, it is most certain that these effects must proceed from the action of some force or other.

But our purpose is only to trace out the quantity and properties of this force from the phenomena (p. 218), and to apply what we discover in some simple cases as principles, by which, in a mathematical way, we may estimate the effects thereof in more involved cases: for it would be endless and impossible to bring every particular to direct and immediate observation.

We said, in a mathematical way, to avoid all questions about the nature or quality of this force, which we would not be understood to determine by any hypothesis; and therefore call it by the general name of a centripetal force, as it is a force which is directed towards some centre; and as it regards more particularly a body in that centre, we call it circum-solar, circum-terrestrial, circum-jovial; and in like manner in respect of other central bodies.

That by means of centripetal forces the planets may be retained in certain orbits, we may easily understand, if we consider the motions of projectiles (p. 75, 76, 77); for a stone projected is by the pressure of its own weight forced out of the rectilinear path, which by the projection alone it should have pursued, and made to describe a curve line in the air; and through that crooked way is at last brought down to the ground; and the greater the velocity is with which it is projected, the farther it goes before it falls to the earth. We may therefore suppose the velocity to be so increased, that it would describe an arc of 1, 2, 5, 10, 100. 1000 miles before it arrived at the earth, till at last, exceeding the limits of the earth, it should pass quite by without touching it.

Let AFB represent the surface of the earth, C its centre, VD, VE, VF, the curve lines which a body would describe, if projected in an horizontal direction from the top of an high mountain successively "with more and

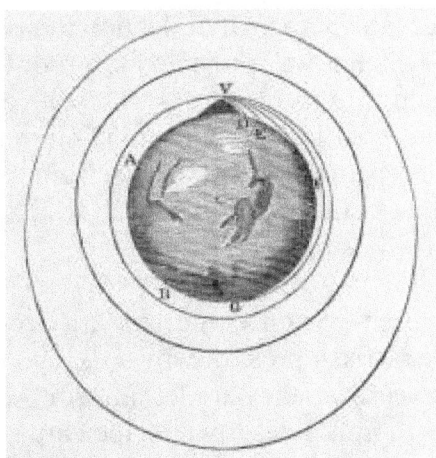

more velocity; and, because the celestial motions are scarcely retarded by the little or no resistance of the spaces in which they are performed, to keep up the parity of cases, let us suppose either that there is no air about the earth, or at least that it is endowed with little or no power of resisting; and for the same reason that the body projected with a less velocity describes the lesser arc VD, and with a greater velocity the greater arc VE. and, augmenting the velocity, it goes farther and farther to F and G, if the velocity was still more and more augmented, it would reach at last quite beyond the circumference of the earth, and return to the mountain from which it was projected.

And since the areas which by this motion it describes by a radius drawn to the centre of the earth are (by Prop. 1, Book 1, Princip. Math.) proportional to the times in which they are described, its velocity, when it returns to the mountain, will be no less than it was at first; and, retaining the same velocity, it will describe the same curve over and over, by the same law

But if we now imagine bodies to be projected in the directions of lines parallel to the horizon from greater heights, as of 5, 10, 100, 1000, or more miles, or rather as many semi-diameters of the earth, those bodies, according to their different velocity, and the different force of gravity in different heights, will describe arcs either concentric with the earth, or variously eccentric, and go on revolving through the heavens in those trajectories, just as the planets do in their orbs.

As when a stone is projected obliquely, that is, any way but in the perpendicular direction, the perpetual deflection thereof towards the earth from the right line in which it was projected is a proof of its gravitation to the earth, no less certain than its direct descent when only suffered to fall freely from rest; so the deviation of bodies moving in free spaces from rectilinear paths, and perpetual deflection therefrom towards any place, is a sure indication of the existence of some force which from all quarters impels those bodies towards that place.

And as, from the supposed existence of gravity, it necessarily follows that all bodies about the earth must press downwards, and therefore must either descend directly to the earth, if they are let fall from rest, or at least perpetually deviate from right lines towards the earth, if they are projected obliquely; so from the supposed existence of a force directed to any centre, it will follow, by the like necessity, that all bodies upon which this force acts mast either descend directly to that centre, or at least deviate perpetually towards it from right lines, if otherwise they should have moved obliquely in these right lines.

And how from the motions given we may infer the forces, or from the forces given we may determine the motions, is shewn in the two first Books of our Principles of Philosophy.

If the earth is supposed to stand still, and the fixed stars to be revolved in free spaces in the space of 24 hours, it is certain the forces by which the fixed stars are retained in their orbs are not directed to the earth, but to the centres of the several orbs, that is, of the several parallel circles, which the fixed stars, declining to one side and the other from the equator, describe daily; also that by radii drawn to the centres of those orbs the fixed stars describe areas exactly proportional to the times of description. Then, because the periodic times are equal (by Cor. Ill, Prop. IV, Book 1), it follows that the centripetal forces are as the radii of the several orbs, and that they will perpetually revolve in the same orbs. And the like consequences may be drawn from the supposed diurnal motion of the planets.

That forces should be directed to no body on which they physically de pend, but to innumerable imaginary points in the axis of the earth, is an hypothesis too incongruous. It is more incongruous still that those forces should increase exactly in proportion of the distances from this axis; for this is an indication of an increase to immensity, or rather to infinity; whereas the forces of natural things commonly decrease in receding from the fountain from which they flow. But, what is yet more absurd, neither are the areas described by the same star proportional to the times, nor are its revolutions performed in the same orb; for as the star recedes from the neighbouring pole, both areas and orb increase; and from the increase of the urea it is demonstrated that the forces are not directed to the axis of the earth. And this difficulty (Cor. 1, Prop. II) arises from the twofold motion that is observed in the fixed stars, one diurnal round the axis of the earth, the other exceedingly slow round the axis of the ecliptic. And the explication thereof requires a composition of forces so perplexed and so variable, that it is hardly to be reconciled with any physical theory.

That there are centripetal forces actually directed to the bodies of the sun, of the earth, and other planets, I thus infer.

The moon revolves about our earth, and by radii drawn to its centre (p. 390) describes areas nearly proportional to the times in which they are described, as is evident from its velocity compared with its apparent diameter; for its motion is slower when its diameter is less (and therefore its distance greater), and its motion is swifter when its diameter is greater.

The revolutions of the satellites of Jupiter about that planet are more regular (p. 386): for they describe circles concentric with Jupiter by equable motions, as exactly as our senses can distinguish.

And so the satellites of Saturn are revolved about this planet with motions nearly (p. 387) circular and equable, scarcely disturbed by any eccentricity hitherto observed.

That Venus and Mercury are revolved about the sun, is demonstrable from their moon-like appearances (p. 388). When they shine with a full face, they are in those parts of their orbs which in respect of the earth lie beyond the sun; when they appear half full, they are in those parts which are over against the sun; when horned, in those parts which lie between the earth and the sun; and sometimes they pass over the sun's disk, when directly interposed between the earth and the sun.

And Venus, with a motion almost uniform, describes an orb nearly circular and concentric with the sun.

But Mercury, with a more eccentric motion, makes remarkable approaches to the sun, and goes off again by turns; but it is always swifter as it is near to the sun, and therefore by a radius drawn to the sun still describes areas proportional to the times.

Lastly, that the earth describes about the sun, or the sun about the earth, by a radius from the one to the other, areas exactly proportional to the times, is demonstrable from the apparent diameter of the sun com pared with its apparent motion.

These are astronomical experiments; from which it follows, by Prop. I,II, III, in the first Book of our Principles, and their Corollaries (p. 213, 214), that there are centripetal forces actually directed (either accurately or without considerable error) to the centres of the earth, of Jupiter, of Saturn, and of the sun. In Mercury, Venus, Mars, and the lesser planets, where experiments are wanting, the arguments from analogy must be allowed in their place.

That those forces (p. 212, 213, 214) decrease in the duplicate proportion of the distances from the centre of every planet, appears by Cor. VI, Prop. IV, Book 1; for the periodic times of the satellites of Jupiter are one to another (p. 386, 387) in the sesquiplicate proportion of their distances from the centre of this planet.

This proportion has been long ago observed in those satellites; and Mr. Flamsted, who had often measured their distances from Jupiter by the micrometer, and by the eclipses of the satellites, wrote to me, that it holds to all the accuracy that possibly can be discerned by our senses. And he sent me the dimensions of their orbits taken by the micrometer, and reduced to the mean distance of Jupiter from the earth, or from the sun, together with the times of their revolutions, as follows:

The greatest elongation of the satelites from the centre of Jupiter as seen from the sun.					The periodic times of their revolutions.			
	′	″		″	d	h	′	″
1st	1	48	or	108	1	18	28	36
2d	3	01	or	181	3	13	17	54
3d	4	46	or	286	7	03	59	36
4th	8	13½	or	493½	16	18	5	13

Whence the sesquiplicate proportion may be easily seen. For example; the 16d 18h 05′ 13″ is to the time 1d.18h.28′ 36″ as 493½″ × √493½″ to 108×√108″, neglecting those small fractions which, in observing, cannot be certainly determined.

Before the invention of the micrometer, the same distances were determined by semi-diameters of Jupiter thus:

Distance of the	1st	2d	3d	4th
By Galileo, . . .	6	10	16	28
" Simon Marius .	6	10	16	26
" Cassini . . .	5	8	13	23
" Borelli, more exactly . . .	5 $^2/_3$	8 $^2/_3$	14	24 $^2/_3$

After the invention of the micrometer:

By Townley . . .	5,51	8,78	13,47	24,72
" Flamsted . . .	5,31	8;85	13.98	24,23
More accurately by the eclipses . .	5,578	8.876	14.159	24,903

And the periodic times of those satellites, by the observations of Mr. Flamsted, are 1d. 18h. 28′ 36″ | 3d. 13h. 17′ 54″ | 7d. 3h. 59′ 36″ | 16d. 18h. 5′ 13″ as above.

And the distances thence computed are 5,578 | 8,878 | 14,168 | 24,968, accurately agreeing with the distances by observation.

Cassini assures us (p. 388, 389) that the same proportion is observed in the circum-saturnal planets. But a longer course of observations is required before we can have a certain and accurate theory of those planets.

In the circum-solar planets, Mercury and Venus, the same proportion holds with great accuracy, according to the dimensions of their orbs, as determined by the observations of the best astronomers.

That Mars is revolved about the sun is demonstrated from the phases which it shews, and the proportion of its apparent diameters (p. 388, 389, and 390); for from its appearing fall near conjunction with the sun, and gibbous in its quadratures, it is certain that it surrounds the sun.

And since its diameter appears about five times greater when in opposition to the sun than when in conjunction therewith, and its distance from the earth is reciprocally as its apparent diameter, that distance will be about five times less when in opposition to than when in conjunction with the sun; but in both cases its distance from the sun will be nearly about the same with the distance which is inferred from its gibbous appearance in the quadratures. And as it encompasses the sun at almost equal distances, but in respect of the earth is very unequally distant, so by radii drawn to the sun it describes areas nearly uniform; but by radii drawn to the earth, it is sometimes swift, sometimes stationary, and sometimes retrograde.

That Jupiter, in a higher orb than Mars, is likewise revolved about the sun, with a motion nearly equable, as well in distance as in the areas described, I infer thus.

Mr. Flamsted assured me, by letters, that all the eclipses of the inner most satellite which hitherto have been well observed do agree with his theory so nearly, as never to differ therefrom by two minutes of time; that in the outmost the error is little greater; in the outmost but one, scarcely three times greater; that in the innermost but one the difference is indeed much greater, yet so as to agree as nearly with his computation? as the moon does with the common tables; and that he computes those eclipses only from the mean motions corrected by the equation of light discovered and introduced by Mr. Rower. Supposing, then, that the theory differs by a less error than that of 2' from the motion of the outmost satellite as hitherto described, and taking as the periodic time 16d. 18h. 5' 13" to 2 in time, so is the whole circle or 360 to the arc 1' 48", the error of Mr. Flamsted's computation, reduced to the satellite's orbit, will be less than 1' 48"; that is, the longitude of the satellite, as seen from the centre of Jupiter, will be determined with a less error than 1' 48". But when the satellite is in the middle of the shadow, that longitude is the same with the heliocentric longitude of Jupiter; and, therefore, the hypothesis which Mr. Flamsted follows, viz., the Copernican, as improved by Kepler, and fas to the motion of Jupiter) lately corrected by himself, rightly represents that longitude within a less error than 1' 48"; but by this longitude, together with the geocentric longitude, which is always easily found, the distance of Jupiter from the sun is determined; which must, therefore, be the very same with that which the hypothesis exhibits. For that greatest error of 1' 48" that can happen in the heliocentric longitude is almost insensible, and quite to be neglected, and perhaps may arise from some yet undiscovered eccentricity of the satellite: but since both longitude and distance are rightly determined, it follows of necessity that Jupiter, by radii drawn to the sun, describes areas so conditioned as the hypothesis requires, that is, proportional to the times.

And the same thing may be concluded of Saturn from his satellite, by the observations of Mr. Huygens and Dr. Halley ; though a longer series of observations is yet wanting to confirm the thing, and to bring it under a sufficiently exact computation.

For if Jupiter was viewed from the sun, it would never appear retrograde nor stationary, as it is seen sometimes from the earth, but always to go forward with a motion nearly uniform (p. 389). And from the very great inequality of its apparent geocentric motion, we infer (by Prop. III Cor. IV) that the force by which Jupiter is turned out of a rectilinear course, and made to revolve in an orb, is not directed to the centre of the earth. And the same argument holds good in Mars and in Saturn. Another centre of these forces is therefore to be looked for (by Prop. II and III, and the Corollaries of the latter), about which the areas described by radii intervening may be equable; and that this is the sun, we have proved already in Mars and Saturn nearly, but accurately enough in Jupiter. It may be alledged that the sun and planets are impelled by some other force equally and in the direction of parallel lines; but by such a force (by Cor. VI of the Laws of Motion) no change would happen in the situation of the planets one to another, nor any sensible effect follow: but our business is with the causes of sensible effects. Let us, therefore, neglect every such force as imaginary and precarious, and of no use in the phenomena of the heavens; and the whole remaining force by which Jupiter is impelled will be directed (by Prop. Ill, Cor. I) to the centre of the sun.

The distances of the planets from the sun come out the same, whether, with Tycho, we place the earth in the centre of the system, or the sun with Copernicus: and we have already proved that these distances are true in Jupiter.

Kepler and Bullialdus have, with great care (p. 388), determined the distances of the planets from the sun; and hence it is that their tables agree best with the heavens. And in all the planets, in Jupiter and Mars, in Saturn and the earth, as well as in Venus and Mercury, the cubes of their distances are as the squares of their periodic times; and therefore (by Cor. VI, Prop. IV) the centripetal circum-solar force throughout all the planetary regions decreases in the duplicate proportion of the distances from the sun. In examining this proportion, we are to use the mean distances, or the transverse semi-axes of the orbits (by Prop. XV), and to neglect those little fractions, which, in denning the orbits, may have arisen from the in sensible errors of observation, or may be ascribed to other causes which we shall afterwards explain. And thus we shall always find the said proportion to hold exactly; for the distances of Saturn, Jupiter, Mars, the Earth, Venus, and Mercury, from the sun, drawn from the observations of astronomers, are, according to the computation of Kepler, as the numbers 951000, 519650, 152350, 100000, 72400, 38806; by the computation of Bullialdus, as the numbers 954198, 522520, 152350, 100000, 72393, 38585; and from the periodic times they come out 953806, 520116, 152399, 100000, 72333, 38710. Their distances, according to Kepler and Bullialdus, scarcely differ by any sensible quantity, and where they differ most the distances drawn from the periodic times, fall in between them.

That the circum-terrestrial force likewise decreases in the duplicate proportion of the distances, I infer thus.

The mean distance of the moon from the centre of the earth, is, in semi-diameters of the earth, according to Ptolemy, Kepler in his Ephemerides, Bullialdus, Hevelius, and Ricciolus, 59; according to Flamsted, 59 1/2; according to Tycho, 56 1/2; to Vendelin, 60; to Copernicus, 60 1/3: to Kircher, 62 1/2 (p . 391, 392, 393).

But Tycho, and all that follow his tables of refraction, making the refractions of the sun and moon (altogether against the nature of light) to exceed those of the fixed stars, and that by about four or five minutes in the horizon, did thereby augment the horizontal parallax of the moon by about the like number of minutes; that is, by about the 12th or 15th part of the whole parallax. Correct this error, and the distance will be come 60 or 61 semi-diameters of the earth, nearly agreeing with what others have determined.

Let us, then, assume the mean distance of the moon 60 semi-diameters of the earth, and its periodic time in respect of the fixed stars 27d. 7h. 43', as astronomers have determined it. And (by Cor. VI, Prop. IV) a body revolved in our air, near the surface of the earth supposed at rest, by means of a centripetal force which should be to the same force at the distance of the moon in the reciprocal duplicate proportion of the distances from the centre of the earth, that is, as 3600 to 1, would (secluding the resistance of the air) complete a revolution in 1h. 24' 27".

Suppose the circumference of the earth to be 123249600 Paris feet; as has been determined by the late mensuration of the French (vide p. 406); then the same body, deprived of its circular motion, and falling by the impulse of the same centripetal force as before, would, in one second of time, describe 15 1/12 Paris feet.

This we infer by a calculus formed upon Prop. XXXVI, and it agrees with what we observe in all bodies about the earth. For by the experiments of pendulums, and a computation raised thereon, Mr. Huygens has demonstrated that bodies falling by all that centripetal force with which (of whatever nature it is) they are impelled near the surface of the earth, do, in one second of time, describe 15 1/12 Paris feet.

But if the earth is supposed to move, the earth and moon together (by Cor. IV of the Laws of Motion, and Prop. LVII) will be revolved about their common centre of gravity. Ana the moon (by Prop. LX) will in the same periodic time, 27d. 7h. 43′, with the same circum terrestrial force diminished in the duplicate proportion of the distance, describe an orbit whose semi-diameter is to the semi-diameter of the former orbit, that is, to 60 semi-diameters of the earth, as the sum of both the bodies of the earth and moon to the first of two mean proportionals between this sum and the body of the earth; that is, if we suppose the moon (on account of its mean apparent diameter 31 1/2′) to be about 1/42 of the earth, as 43 to $\sqrt[3]{(42 + 43^2)}$, or as about 128 to 127. And therefore the semi-diameter of the orbit, that is, the distance between the centres of the moon and earth, will in this case be 60 1/2 semi-diameters of the earth, almost the same with that assigned by Copernicus, which the Tychonic observations by no means disprove; and, therefore, the duplicate proportion of the decrement of the force holds good in this distance. I have neglected the increment of the orbit which arises from the action of the sun as inconsiderable; but if that is subducted, the true distance will remain about 60 4/9 semi-diameters of the earth.

But farther (p. 390); this proportion of the decrement of the forces is confirmed from the eccentricity of the planets, and the very slow motion of their apses; for (by the Corollaries of Prop. XLV) in no other proportion could the circum-solar planets once in every revolution descend to their least and once ascend to their greatest distance from the sun, and the places of those distances remain immoveable. A small error from the duplicate proportion would produce a motion of the apses considerable in every revolution, but in many enormous.

But now, after innumerable revolutions, hardly any such motion has been perceived in the orbs of the circum-solar planets. Some astronomers affirm that there is no such motion; others reckon it no greater than what may easily arise from the causes hereafter to be assigned, and is of no moment in the present question.

We may even neglect the motion of the moon's apsis (p. 390, 391), which is far greater than in the circum-solar planets, amounting in every revolution to three degrees; and from this motion it is demonstrable that the circum-terrestrial force decreases in no less than the duplicate, but far less than the triplicate proportion of the distance; for if the duplicate proportion was gradually changed into the triplicate, the motion of the

14

apsis would thereby increase to infinity; and, therefore, by a very small mutation, would exceed the motion of the moon's apsis. This slow motion arises from the action of the circum-solar force, as we shall afterwards explain. But, secluding this cause, the apsis or apogeon of the moon will be fixed, and the duplicate proportion of the decrease of the circum-terrestrial force in different distances from the earth will accurately take place.

Now that this proportion has been established, we may compare the forces of the several planets among themselves (p. 391).

In the mean distance of Jupiter from the earth, the greatest elongation of the outmost satellite from Jupiter's centre (by the observations of Mr. Flamsted] is 8′ 13″; and therefore the distance of the satellite from the centre of Jupiter is to the mean distance of Jupiter from the centre of the sun as 124 to 52012, but to the mean distance of Venus from the centre of the sun as 124 to 7234; and their periodic times are $16^3/_4$d. and $224^2/_3$d; and from hence (according to Cor. II, Prop. IV), dividing the distances by the squares of the times, we infer that the force by which the satellite is impelled towards Jupiter is to the force by which Venus is impelled to wards the sun as 442 to 143; and if we diminish the force by which the satellite is impelled in the duplicate proportion of the distance 124 to 7234, we shall have the circum-jovial force in the distance of Venus from the sun to the circum-solar force by which Venus is impelled as $^{13}/_{100}$ to 143, or as 1 to 1100; wherefore at equal distances the circum-solar force is 1100 times greater than the circum-jovial.

And, by the like computation, from the periodic time of the satellite of Saturn 15d. 22h. and its greatest elongation from Saturn, while that planet is in its mean distance from us, 3′ 20″, it follows that the distance of this satellite from Saturn's centre is to the distance of Venus from the sun as $92^2/_3$ to 7234; and from thence that the absolute circum-solar force is 2360 times greater than the absolute circum-saturnal.

From the regularity of the heliocentric and irregularity of the geocentric motions of Venus, of Jupiter, and the other planets, it is evident (by Cor. IV, Prop. Ill) that the circum-terrestrial force, compared with the circum-solar, is very small.

Ricciolus and Vendelin have severally tried to determine the sun's parallax from the moon's dichotomies observed by the telescope, and they agree that it does not exceed half a minute.

Kepler, from Tycho's observations and his own, found the parallax of Mars insensible, even in opposition to the sun, when that parallax is some thing greater than the sun's.

Flamsted attempted the same parallax with the micrometer in the perigeon position of Mars, but never found it above 25″; and thence concluded the sun's parallax at most 10″.

Whence it follows that the distance of the moon from the earth bears no greater proportion to the distance of the earth from the sun than 29 to 10000: nor to the distance of Venus from the sun than 29 to 7233.

From which distances, together with the periodic times, by the method above explained, it is easy to infer that the absolute circum-soiar force is greater than the absolute circum-terrestrial force at least 229400 times.

And though we were only certain, from the observations of Ricciolus and Vendelin, that the sun's parallax was less than half a minute, yet from this it will follow that the absolute circum-solar force exceeds the absolute circum-terrestrial force 8500 times.

By the like computations I happened to discover an analogy, that is observed between the forces and the bodies of the planets; but, before I explain this analogy, the apparent diameters of the planets in their mean distances from the earth must be first determined.

Mr. Flamsted (p. 387), by the micrometer, measured the diameter of Jupiter 40″ or 41″; the diameter of Saturn's ring 50″; and the diameter of the sun about 32 13″ (p. 387).

But the diameter of Saturn is to the diameter of the ring, according to Mr. Huygens and Dr. Halley, as 4 to 9; according to Gulletius, as 4 to 10; and according to Hooke (by a telescope of 60 feet), as 5 to 12. And from the mean proportion, 5 to 12, the diameter of Saturn's body is inferred about 21″.

Such as we have said are the apparent magnitudes; but. because of the unequal refrangibility of light, all lucid points are dilated by the telescope, and in the focus of the object-glass possess a circular space whose breadth is about the 50th part of the aperture of the glass.

It is true, that towards the circumference the light is so rare as hardly to move the sense; but towards the middle, where it is of greater density, and is sensible enough, it makes a small lucid circle, whose breadth varies according to the splendor of the lucid point, but is generally about the 3d, or 4th, or 5th part of the breadth of the whole.

Let ABD represent the circle of the whole light; PQ the small circle of the denser and clearer light; C the centre of both; CA, CB, semi-diameters of the greater circle containing a right angle at C; ACBE the square comprehended under these semi-diameters; AB the diagonal of that square; EGH an hyperbola with the centre C and asymptotes CA, CB PG a perpendicular erected from any point P of the line BC, and meeting the hyperbola in G, and the right lines AB, AE, in K and F: and the density of the light in any place P, will, by my computation, be as the line FG, and therefore at the centre infinite, but near the circumference very small. And the whole light within the small circle PQ, is to the without as the area of the quadrilateral figure CAKP to the triangle

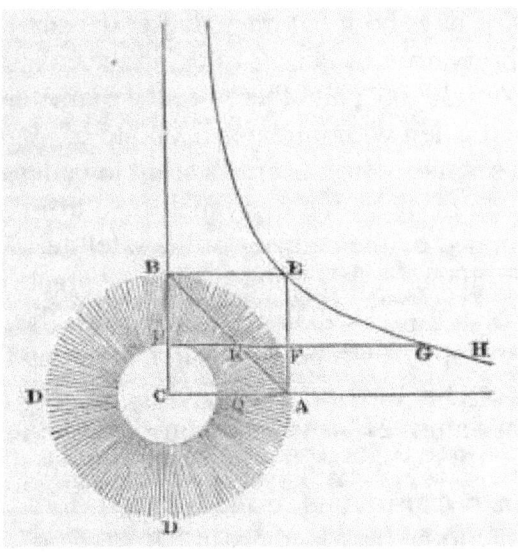

PKB. And we are to understand the small circle PQ, to be there terminated, where FG, the density of the light, begins to be less than what is required to move the sense.

Hence it was, that, at the distance of 191,382 feet, a fire of 3 feet in diameter, through a telescope of 3 feet, appeared to Mr. Picart of 8″ in breadth, when it should have appeared only of 3″ 14‴; and hence it is that the brighter fixed stars appear through the telescope as of 5″ or 6″ in diameter, and that with a good full light; but with a fainter light they appear to run out to a greater breadth. Hence, likewise, it was that Hevelius, by diminishing the aperture of the telescope, did cut off a great part of the light towards the circumference, and brought the disk of the star to be more distinctly defined, which, though hereby diminished, did yet appear as of 5″ or 6″ in diameter. But Mr. Huygens, only by clouding the eye-glass with a little smoke, did so effectually extinguish this scattered light, that the fixed stars appeared as mere points, void of all sensible breadth. Hence also it was that Mr. Huygens, from the breadth of bodies interposed to intercept the whole light of the planets, reckoned their diameters greater than others have measured them by the micrometer: for the scattered light, which could not be seen before for the stronger light of the planet, when the planet is hid, appears every way farther spread. Lastly, from hence it is that the planets appear so small in the disk of the sun, being lessened by the dilated light. For to Hevelius, Galletius, and Dr. Halley, Mercury did not seem to exceed 12″ or 15″; and Venus appeared to Mr. Crabtrie only 1′ 3″; to Horrox but 1′ 12″; though by the mensurations of Hevelius and Hugenius without the sun's disk, it ought to have been seen at least 1′ 24″. Thus the apparent diameter of the moon, which in 1684, a few days both before and after the sun's eclipse, was measured at the observatory of Paris 31 30″, in the eclipse itself did not seem to exceed 30′ or 30′ 05″; and therefore the diameters of the planets are to be diminished when without the sun, and to be augmented when within it, by some seconds. But the errors seem to be less than usual in the mensurations that are made by the micrometer. So from the diameter of the shadow, determined by the eclipses of the satellites, Mr. Flamsted found that the semi-diameter of Jupiter was to the greatest elongation of the outmost satellite as 1 to 24,903. Wherefore since that elongation is 8′ 13″, the diameter of Jupiter will be 39½″; and, rejecting the scattered light, the diameter found by the micrometer 40″ or 41″ will be reduced to 39½″; and the diameter of Saturn 21″ is to be diminished by the like correction, and to be reckoned 20″, or something less. But (if I am not mistaken) the diameter of the sun, because of its stronger light, is to be diminished something more, and to be reckoned about 32′, or 32′ 6″.

That bodies so different in magnitude should come so near to an analogy with their forces, is not without some mystery (p. 400).

It may be that the remoter planets, for want of heat, have not those metallic substances and ponderous minerals with which our earth abounds; and that the bodies of Venus and Mercury, as they are more exposed to the sun's heat, are also harder baked, and more compact.

For, from the experiment of the burning-glass, we see that the heat in creases with the density of light; and this density increases in the reciprocal duplicate proportion of the distance from the sun; from whence the sun's heat in Mercury is proved to be sevenfold its heat in our summer seasons. But with this heat our water boils; and those heavy fluids, quick silver and the spirit of vitriol, gently evaporate, as I have tried by the thermometer; and therefore there can be no fluids in Mercury but what are heavy, and able to bear a great heat, and from which substances of great density may be nourished.

And why not, if God has placed different bodies at different distances from the sun, so as the denser bodies always possess the nearer places, and each body enjoys a degree of heat suitable to its condition, and proper for its nourishment? From this consideration it will best appear that the weights of all the planets are one to another as their forces.

But I should be glad the diameters of the planets were more accurately measured; and that may be done, if a lamp, set at a great distance, is made to shine through a circular hole, and both the hole and the light of the lamp are so diminished that the spectrum may appear through the telescope just like the planet, and may be defined by the same measure: then the diameter of the hole will be to its distance from the objective glass as the true diameter of the planet to its distance from us. The light of the lamp may be diminished by the interposition either of pieces of cloth, or of smoked glass.

Of kin to the analogy we have been describing, there is another observed between the forces and the bodies attracted (p. 395, 396, 397). Since the action of the centripetal force upon the planets decreases in the duplicate proportion of the distance, and the periodic time increases in the sesquiplicate thereof, it is evident that the actions of the centripetal force, and therefore the periodic times, would be equal in equal planets at equal distances from the sun; and in equal distances of unequal planets the total actions of the centripetal force would be as the bodies of the planets; for if the actions were not proportional to the bodies to be moved, they could not equally retract these bodies from the tangents of their orbs in equal times:

nor could the motions of the satellites of Jupiter be so regular, if it was not that the circum-solar force was equally exerted upon Jupiter and all its satellites in proportion of their several weights. And the same thing is to be said of Saturn in respect of its satellites, and of our earth in respect of the moon, as appears from Cor. II and III, Prop. LXV. Arid, therefore, at equal distances, the actions of the centripetal force are equal upon all the planets in proportion of their bodies, or of the quantities of matter in their several bodies; and for the same reason must be the same upon all the particles of the same size of which the planet is composed; for if the action was greater upon some sort of particles than upon others than in proportion to their quantity of matter, it would be also greater or less upon the whole planets not in proportion to the quantity only, but like wise of the sort of the matter more copiously found in one and more sparingly in another.

In such bodies as are found on our earth of very different sorts, I examined this analogy with great accuracy (p. 343, 344).

If the action of the circum-terrestrial force is proportional to the bodies to be moved, it will (by the Second Law of Motion) move them with equal velocity in equal times, and will make all bodies let fall to descend through equal spaces in equal times, and all bodies hung by equal threads to vibrate in equal times. If the action of the force was greater, the times would be less; if that was less, these would be greater.

But it has been long ago observed by others, that (allowance being made for the small resistance of the air) all bodies descend through equal spaces in equal times; and, by the help of pendulums, that equality of times may be distinguished to great exactness.

I tried the thing in gold, silver, lead, glass, sand, common salt wood, water, and wheat. I provided two equal wooden boxes. I filled the one with wood, and suspended an equal weight of gold (as exactly as I could) in the centre of oscillation of the other. The boxes, hung by equal threads of 11 feet, made a couple of pendulums perfectly equal in weight and figure, and equally exposed to the resistance of the air: and, placing the one by the other, I observed them to play together forwards and backwards for a long while, with equal vibrations. And therefore (by Cor. 1 and VI, Prop. XXIV. Book II) the quantity of matter in the gold was to the quantity of matter in the wood as the action of the motive force upon all the gold to the action of the same upon all the wood; that is, as the weight of the one to the weight of the other.

And by these experiments, in bodies of the same weight, could have dis covered a difference of matter less than the thousandth part of the whole.

Since the action of the centripetal force upon the bodies attracted is, at equal distances, proportional to the quantities of matter in those bodies, reason requires that it should be also proportional to the quantity of matter in the body attracting.

For all action is mutual, and (p. 83, 93. by the Third Law of Motion) makes the bodies mutually to approach one to the other, and therefore must be the same in both bodies. It is true that we may consider one body as attracting, another as attracted; but this distinction is more mathematical than natural. The attraction is really common of either to other, and therefore of the same kind in both.

And hence it is that the attractive force is found in both. The sun at tracts Jupiter and the other planets; Jupiter attracts its satellites; and, for the same reason, the satellites act as well one upon another as upon Jupiter, and all the planets mutually one upon another.

And though the mutual actions of two planets may be distinguished and considered as two, by which each attracts the other, yet, as those actions are intermediate, they do not make two but one operation between two terms. Two bodies may be mutually attracted each to the other by the contraction of a cord interposed. There is a double cause of action, to wit, the disposition of both bodies, as well as a double action in so far as the action is considered as upon two bodies; but as betwixt two bodies it is but one single one. It is not one action by which the sun attracts Jupiter, and another by which Jupiter attracts the sun; but it is one action by which the sun and Jupiter mutually endeavour to approach each the other. By the action with which the sun attracts Jupiter, Jupiter and the sun endeavours to come nearer together (by the Third Law of Motion); and by the action with which Jupiter attracts the sun, likewise Jupiter and the sun endeavor to come nearer together. But the sun is not attracted towards Jupiter by a twofold action, nor Jupiter by a twofold action towards the sun; but it is one single intermediate action, by which both approach nearer together.

Thus iron draws the load-stone (p. 93), as well as the load-stone draws the iron: for all iron in the neighbourhood of the load-stone draws other iron. But the action betwixt the load-stone and iron is single, and is considered as single by the philosophers. The action of iron upon the load-stone, is, indeed, the action of the load-stone betwixt itself and the iron, by which both endeavour to come nearer together: and so it manifestly appears; for if you remove the load-stone, the whole force of the iron almost ceases.

Tn this sense it is that we are to conceive one single action to be exerted betwixt two planets, arising from the conspiring natures of both: and this action standing in the same relation to both, if it is proportional to the quantity of matter in the one, it will be also proportional to the quantity of matter in the other.

Perhaps it may be objected, that, according to this philosophy (p. 398), all bodies should mutually attract one another, contrary to the evidence of experiments in terrestrial bodies; but I answer, that the experiments in terrestrial bodies come to no account; for the attraction of homogeneous spheres near their surfaces are (by Prop. LXXII) as their diameters. Whence a sphere of one foot in diameter, and of a like nature to the earth, would attract a small body placed near its surface with a force 20000000 times less than the earth would do if placed near its surface; but so small a force could produce no sensible effect. If two such spheres were distant but by 1 of an inch, they would not, even in spaces void of

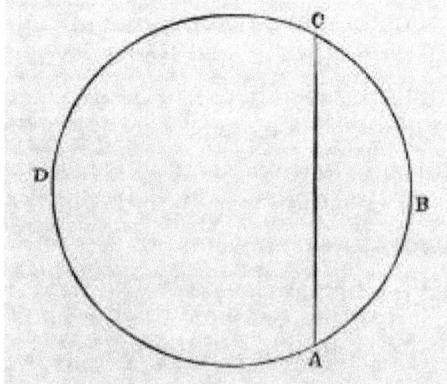

resistance, come together by the force of their mutual attraction in less than a month's time; and less spheres will come together at a rate yet slower, viz. in the proportion of their diameters. Nay, whole mountains will not be sufficient to produce any sensible effect. A mountain of an hemispherical figure, three miles high, and six broad, will not, by its attraction, draw the pendulum two minutes out of the true perpendicular: and it is only in the great bodies of the planets that these forces are to be perceived, unless we may reason about smaller bodies in manner following.

Let ABCD (p. 93) represent the globe of the earth cut by any plane AC into two parts ACB, and A CD. The part ACB bearing upon the part ACD presses it with its whole weight; nor can the part ACD sustain this pressure and continue unmoved, if it is not opposed by an equal contrary pressure. And therefore the parts equally press each other by their weights, that is, equally attract each other, according to the third Law of Motion; and, if separated and let go, would fall towards each other with velocities reciprocally as the bodies. All which we may try and see in the load-stone, whose attracted part does not propel the part attracting, but is only stopped and sustained thereby.

Suppose now that ACB represents some small body on the earth's surface: then, because the mutual attractions of this particle, and of the remaining part ACD of the earth towards each other, are equal, but the attraction of the particle towards the earth (or its weight) is as the matter of the particle (as we have proved by the experiment of the pendulums), the attraction of the earth towards the particle will likewise be as the matter of the particle; and therefore the attractive forces of all terrestrial bodies will be as their several quantities of matter.

The forces (p. 396), which are as the matter in terrestrial bodies of all forms, and therefore are not mutable with the forms, must be found in all sorts of bodies whatsoever, celestial as well as terrestrial, and be in all proportional to their quantities of matter, because among all there is no difference of substance, but of modes and forms only. But in the celestial bodies the same thing is likewise proved thus. We have shewn that the action of the circum-solar force upon all the planets (reduced to equal distances) is as the matter of the planets; that the action of the circum-jovial force upon the satellites of Jupiter observes the same law; and the same thing is to be said of the attraction of all the planets towards every planet: but thence it follows (by Prop. LXIX) that their attractive forces are as their several quantities of matter.

As the parts of the earth mutually attract one another, so do those of all the planets. If Jupiter and its satellites were brought together, and formed into one globe, without doubt they would continue mutually to attract one another as before. And, on the other hand, if the body of Jupiter was broke into more globes, to be sure, these would no less attract one another than they do the satellites now. From these attractions it is that the bodies of the earth and all the planets effect a spherical figure, and their parts cohere, and are not dispersed through the aether. But we have before proved that these forces arise from the universal nature of matter (p. 398), and that, therefore, the force of any whole globe is made up of the several forces of all its parts. And from thence it follows (by Cor. III, Prop. LXXIV) that the force of every particle decreases in the duplicate proportion of the distance from that particle; and (by Prop. LXXIII and LXXV) that the force of an entire globe, reckoning from the surface outwards, decreases in the duplicate, but, reckoning inwards, in the simple proportion of the distances from the centres, if the matter of the globe be uniform. And though the matter of the globe, reckoning from the centre towards the surface, is not uniform (p. 398, 399), yet the decrease in the duplicate proportion of the distance outwards would (by Prop. LXXVI) take place, provided that difformity is similar in places round about at equal distances from the centre. And two such globes will (by the same Proposition) attract one the other with a force decreasing in the duplicate proportion of the distance between, their centres.

Wherefore the absolute force of every globe is as the quantity of matter which the globe contains; but the motive force by which every globe is attracted towards another, and which, in terrestrial bodies, we commonly call their weight, is as the content under the quantities of matter in both globes applied to the square of the distance between their centres (by Cor. IV, Prop. LXXVI), to which force the quantity of motion, by which each globe in a given time will be carried towards the other, is proportional. And the accelerative force, by which every globe according to its quantity of matter is attracted towards another, is as the quantity of matter in that other globe applied to the square of the distance between the centres of the two (by Cor. II, Prop. LXXVI): to which force, the velocity by which the attracted globe will, in a given time, be carried towards the other is proportional. And from these principles well understood, it will be now easy to determine the motions of the celestial bodies among themselves.

From comparing the forces of the planets one with another, we have above seen that the circum-solar does more than a thousand times exceed all the rest; but by the action of a force so great it is unavoidable but that all bodies within, nay, and far beyond, the bounds of the planetary system must descend directly to the sun, unless by other motions they are impelled towards other parts: nor is our earth to be excluded from the number of such bodies: for certainly the moon is a body of the same nature with the planets, and subject to the same attractions with the other planets, seeing it is by the circum-terrestrial force that it is retained in its orbit. But that the earth and moon are equally attracted towards the sun, we have above proved; we have likewise before proved that all bodies are subject to the said common laws of attraction. Nay, supposing any of those bodies to be deprived of its circular motion about the sun, by having its distance from the sun, we may find (by Prop. XXXVI) in what space of time it would in its descent arrive at the sun; to wit, in half that periodic time in which the body might be revolved at one half of its former distance; or in a space of time that is to the periodic time of the planet as 1 to $4\sqrt{2}$; as that Venus in its descent would arrive at the sun in the space of 40 days, Jupiter in the space of two years and one month, and the earth and moon together in the space of 66 days and 19 hours. But, since no such thing happens, it must needs be, that those bodies are moved towards other parts (p. 75), nor is every motion sufficient for this purpose. To hinder such a descent, a due proportion of velocity is required. And hence depends the force of the argument drawn from the retardation of the motions of the planets. Unless the circum-solar force decreased in the duplicate ratio of their increasing slowness, the excess thereof would force those bodies to descend to the sun; for instance, if the motion (*cæteris paribus*) was retarded by one half, the planet would be retained in its orb by one fourth of the former circum-solar force, and by the excess of the other three fourths would descend to the sun. And therefore the planets (Saturn, Jupiter, Mars, Venus, and Mercury) are not really retarded in their perigees, nor become really stationary, or regressive with slow motions. All these are but apparent, and the absolute motions, by which the planets continue to revolve in their orbits, are always direct, and nearly equable. But that such motions are performed about the sun, we have already proved; and therefore the sun, as the centre of the absolute motions, is quiescent. For we can by no means allow quiescence to the earth, lest the planets in their perigees should indeed be truly retarded, and become truly stationary and regressive, and so for want of motion should descend to the sun. But farther; since the planets (Venus, Mars, Jupiter, and the rest) by radii drawn to the sun describe regular orbits, and areas (as we have shewn) nearly and to sense proportional to the times, it follows (by Prop. III. and Cor. III, Prop. LXV) that the sun is moved with no notable force, unless perhaps with such as all the planets are equally moved with, according to their several quantities of matter, in parallel lines, and so the whole system is transferred in right lines. Reject that translation of the whole system, and the sun will be almost quiescent in the centre thereof. If the gun was revolved about the earth, and carried the other planets round about itself, the earth ought to attract the sun with a great force, but the cir cum-solar planets with no force producing any sensible effect, which is contrary to Cor. Ill, Prop. LXV. Add to this, that if hitherto the earth, because of the gravitation of its parts, has been placed by most authors in the lowermost region of the universe; now, for better reason, the sun possessed of a centripetal force exceeding our terrestrial gravitation a thousand times and more, ought to be depressed into the lowermost place, and to be held for the centre of the system. And thus the true disposition of the whole system will be more fully and more exactly understood.

Because the fixed stars are quiescent one in respect of another (p. 401, 402), we may consider the sun, earth, and planets, as one system of bodies carried hither and thither by various motions among themselves; and the common centre of gravity of all (by Cor. II of the Laws of Motion) will either be quiescent, or move uniformly forward in a right line: in which case the whole system will likewise move uniformly forward in right lines. But this is an hypothesis hardly to be admitted; and, therefore, setting it aside, that common centre will be quiescent: and from it the sun is never far removed. The common centre of gravity of the sun and Jupiter falls on the surface of the sun; and though all the planets were placed towards the same parts from the sun with Jupiter the common centre of the sun and all of them would scarcely recede twice as far from the sun's centre; and, therefore, though the sun, according to the various situation of the planets, is variously agitated, and always wandering to and fro with a slow motion of libration, yet it never recedes one entire diameter of its own body from the quiescent centre of the whole system. But from the weights of the sun and planets above determined, and the situation of all among them selves, their common centre of gravity may be found; and, this being given, the sun's place to any supposed time may be obtained.

About the sun thus librated the other planets are revolved in elliptic orbits (p 403), and, by radii drawn to the sun, describe areas nearly proportional to the times, as is explained in Prop. LXV. If the sun was quiescent, and the other planets did not act mutually one upon another, their orbits would be elliptic, and the areas exactly proportional to the times (by Prop. XI, and Cor. I, Prop. XIII). But the actions of the planets among themselves, compared with the actions of the sun on the planets, are of no moment, and produce no sensible errors. And those errors are less in revolutions about the sun agitated in the manner but now described than if those revolutions were made about the sun quiescent (by Prop. LXVI, and Cor. Prop. LXVIII), especially if the focus of every orbit is placed in the common centre of gravity of all the lower included planets; viz., the focus of the orbit of Mercury in the centre of the sun: the focus of the orbit of Venus in the common centre of gravity of Mercury and the sun; the focus of the orbit of the earth in the common centre of gravity of Venus, Mercury, and the sun; and so of the rest. And by this means the foci of the orbits of all the planets, except Saturn, will not be sensibly removed from the centre of the sun, nor will the focus of the orbit of Saturn recede sensibly from the common centre of gravity of Jupiter and the sun. And therefore astronomers are not far from the truth, when they reckon the sun's centre the common focus of all the planetary orbits. In Saturn itself the error thence arising does not exceed 1 45 . And if its orbit, by placing the focus thereof in the common centre of gravity of Jupiter and the sun, shall happen to agree better with the phenomena, from thence all that we have said will be farther confirmed.

If the sun was quiescent, and the planets did not act one on another, the aphelions and nodes of their orbits would likewise (by Prop. I, XI, and Cor. Prop. XIII) be quiescent. And the longer axes of their elliptic orbits would (by Prop. XV) be as the cubic roots of the squares of their periodic times: and therefore from the given periodic times would be also given. But those times are to be measured not from the equinoctial points, which are moveable, but from the first star of Aries. Put the semi-axis of the earth's orbit 100000, and the semi-axes of the orbits of Saturn, Jupiter, Mars, Venus, and Mercury, from their periodic times, will come out 953806, 520116, 152399, 72333, 38710 respectively. But from the sun's motion every semi-axis is increased (by Prop. LX) by about one third of the distance of the sun's centre from the common centre of gravity of the sun and planet (p. 405, 406.) And from the actions of the exterior planets on the interior, the periodic times of the interior are something protracted, though scarcely by any sensible quantity; and their aphelions are transferred (by Cor. VI and VII, Prop. LXVI) by very slow motions in consequentia. And on the like account the periodic times of all, especially of the exterior planets, will be prolonged by the actions of the comets, if any such there are, without the orb of Saturn, and the aphelions of all will be thereby carried forwards in consequentia. But from the progress of the aphelions the regress of the nodes follows (by Cor. XI, XIII, Prop. LXVI). And if the plane of the ecliptic is quiescent, the regress of the nodes (by Cor. XVI, Prop. LXVI) will be to the progress of the aphelion in every orbit as the regress of the nodes of the moon's orbit to the progress of its apogeon nearly, that is, as about 10 to 21. But astronomical observations seem to confirm a very slow progress of the aphelions, and a regress of the nodes in respect of the fixed stars. And hence it is probable that there are comets in the regions beyond the planets, which, revolving in very eccentric orbs, quickly fly through their perihelion parts, and, by an exceedingly slow motion in their aphelions, spend almost their whole time in the regions beyond the planets; as we shall afterwards ex plain more at large.

The planets thus revolved about the sun (p. 413, 41.4, 415) may at the same time carry others revolving about themselves as satellites or moons, as appears by Prop. LXVI. But from the action of the sun our moon must move with greater velocity, and, by a radius drawn to the earth, de scribe an area greater for the time; it must have its orbit less curve, and therefore approach nearer to the earth in the syzygies than in the quadra tures, except in so far as the motion of eccentricity hinders those effects.

Per the eccentricity is greatest when the moon's apogeon is in the syzygies, and least when the same is in the quadratures; and hence it is that the perigeon moon is swifter and nearer to us, but the apogeon moon slower and farther from us, in the syzygies than in the quadratures. But farther; the apogeon has a progressive and the nodes a regressive motion, both unequable. For the apogeon is more swiftly progressive in its syzygies, more slowly regressive in its quadratures, and by the excess of its progress above its regress is yearly transferred in consequentia; but the nodes are quiescent in their syzygies, and most swiftly regressive in their quadratures. But farther, still, the greatest latitude of the moon is greater in its quadratures than in its syzygies; and the mean motion swifter in the aphelion of the earth than in its perihelion. More inequalities in the moon's motion have not hitherto been taken notice of by astronomers: but all these follow from our principles in Cor. II, III, IV, V, VI, VII, VIII, IX, X, XI, XII, XIII, Prop. LXVI, and are known really to exist in the heavens. And this may seen in that most ingenious, and if I mistake not, of all, the most accurate, hypothesis of Mr. Horrox, which Mr. Flamsted has fitted to the heavens; but the astronomical hypotheses are to be corrected in the motion of the nodes; for the nodes admit the greatest equation or prosthaphaeresis in their octants, and this inequality is most conspicuous when the moon is in the nodes, and therefore also in the octants; and hence it was that Tycho, and others after him, referred this inequality to the octants of the moon, and made it menstrual; but the reasons by us adduced prove that it ought to be referred to the octants of the nodes, and to be made annual.

Beside those inequalities taken notice of by astronomers (p. 414, 445, 447,) there are yet some others, by which the moon's motions are so disturbed, that hitherto by no law could they be reduced to any certain regulation. For the velocities or horary motions of the apogee and nodes of the moon, and their equations, as well as the difference betwixt the greatest eccentricity in the syzygies and the least in the quadratures, and that inequality which we call the variation, in the progress of the year are augmented and diminished (by Cor. XIV, Prop. LXVI) in the triplicate ratio of the sun's apparent diameter. Beside that, the variation is mutable nearly in the duplicate ratio of the time between the quadratures (by Cor. I and II, Lem. X, and Cor. XVI, Prop. LXVI); and all those inequalities are something greater in that part of the orbit which respects the sun than in the opposite part, but by a difference that is scarcely or not at all perceptible.

By a computation (p. 422), which for brevity's sake I do not describe, I also find that the area which the moon by a radius drawn to the earth describes in the several equal moments of time is nearly as the sum of the number 237 3/10, and versed sine of the double distance of the moon from the nearest quadrature in a circle whose radius is unity; and therefore that the square of the moon's distance from the earth is as that sum divided by the horary motion of the moon. Thus it is when the variation in the octants is in its mean quantity; but if the variation is greater or less, that versed sine must be augmented or diminished in the same ratio. Let astronomers try how exactly the distances thus found will agree with the moon's apparent diameters.

From the motions of our moon we may derive the motions of the moon or satellites of Jupiter and Saturn (p. 413); for the mean motion of the nodes of the outmost satellite of Jupiter is to the mean motion of the nodes of our moon in a proportion compounded of the duplicate proportion of the periodic time of the earth about the sun to the periodic time of Jupiter about the sun, and the simple proportion of the periodic time of the satellite about Jupiter to the periodic time of our moon about the earth (by Cor. XVI, Prop. LXVI): and therefore those nodes, in the space of a hundred years, are carried 8° 24′ backwards, or in *antecedentia*. The mean motions of the nodes of the inner satellites are to the (mean) motion of (the nodes of) the outmost as their periodic times to the periodic time of this, by the same corollary, and are thence given. And the motion of the apsis of every satellite in consequentia is to the motion of its nodes in *antecedentia*, as the motion of the apogee of our moon to the motion of its nodes (by the same Corollary), and is thence given. The greatest equations of the nodes and line of the apses of each satellite are to the greatest equations of the nodes and the line of the apses of the moon respectively as the motion of the nodes and line of the apses of the satellites in the time of one resolution of the first equations to the motion of the nodes and apogeon of the moon in the time of one revolution of the last equations. The variation of a satellite seen from Jupiter is to the variation of our moon in the same proportion as the whole motions of their nodes respectively, during the times in which the satellite and our moon (after parting from) are revolved (again) to the sun, by the same Corollary; and therefore in the outmost satellite the variation does not exceed 5″ 12″. From the small quantity of those inequalities, and the slowness of the motions, it happens that the motions of the satellites are found to be so regular, that the more modern astronomers either deny all motion to the nodes, or affirm them to be very slowly regressive. (P. 404). While the planets are thus revolved in orbits about remote centres, in the mean time they make their several rotations about their proper axes; the sun in 26 days; Jupiter in 9h. 56′; Mars in 24 2/3 h.; Venus in 23h.; and that in planes not much inclined to the plane of the ecliptic, and according to the order of the signs, as astronomers determine from the spots or maculae that by turns present themselves to our sight in their bodies; and there is a like revolution of our earth performed in 24h.; and those motions are neither accelerated nor retarded by the actions of the centripetal forces, as appears by Cor. XXII, Prop. LXVI; and therefore of all others they are the most equable and most fit for the mensuration of time; but those revolutions are to be reckoned equable not from their return to the sun, but to some fixed star: for as the position of the planets to the sun is unequably varied, the revolutions of those planets from sun to sun are rendered unequable.

In like manner is the moon revolved about its axis by a motion most equable in respect of the fixed stars, viz., in 27d. 7h. 43′, that is, in the space of a sidereal month; so that this diurnal motion is equal to the mean motion of the moon in its orbit: upon which account the same face of the moon always respects the centre about which this mean motion is performed, that is, the exterior focus of the moon's orbit nearly; and hence arises a deflection of the moon's face from the earth, sometimes towards the east, and other times towards the west, according to the position of the focus which it respects; and this deflection is equal to the equation of the moon's orbit, or to the difference betwixt its mean and true motions; and this is the moon's libration in longitude: but it is likewise affected with a libration in latitude arising from the inclination of the moon's axis to the plane of the orbit in which the moon is revolved about the earth; for that axis retains the same position to the fixed stars nearly, and hence the poles present themselves to our view by turns, as we may understand from the example of the motion of the earth, whose poles, by reason of the inclination of its axis to the plane of the ecliptic, are by turns illuminated by the sun. To determine exactly the position of the moon's axis to the fixed stars, and the variation of this position, is a problem worthy of an astronomer.

By reason of the diurnal revolutions of the planets, the matter which they contain endeavours to recede from the axis of this motion; and hence the fluid parts rising higher towards the equator than about the poles (p. 405), would lay the solid parts about the equator under water, if those parts did not rise also (p. 405, 409): upon which account the planets are something thicker about the equator than about the poles; and their equinoctial points (p. 413) thence become regressive; and their axes, by a motion of nutation, twice in every revolution, librate towards their ecliptics, and twice return again to their former inclination, as is explained in Cor. XVIII, Prop. LXVI; and hence it is that Jupiter, viewed through very long telescopes, does not appear altogether round (p. 409). but having its diameter that lies parallel to the ecliptic something longer than that which is drawn from north to south.

And from the diurnal motion and the attractions (p. 415, 418) of the Bun and moon our sea ought twice to rise and twice to fall every day, as well lunar as solar (by Cor. XIX, XX, Prop. LXVI), and the greatest height of the water to happen before the sixth hour of either day and after the twelfth hour preceding. By the slowness of the diurnal motion the flood is retracted to the twelfth hour; and by the force of the motion of reciprocation it is protracted and deferred till a time nearer to the sixth hour. But till that time is more certainly determined by the phenomena, choosing the middle between those extremes, why may we not conjecture the greatest height of the water to happen at the third hour? for thus the water will rise all that time in which the force of the luminaries to raise it is greater, and will fall all that time in which their force is less: viz., from the ninth to the third hour when that force is greater, and from the third to the ninth when it is less. The hours I reckon from the appulse of each luminary to the meridian of the place, as well under as above the horizon; and by the hours of the lunar day I understand the twenty-fourth parts of that time which the moon spends before it comes about again by its apparent diurnal motion to the meridian of the place which it left the day before.

But the two motions which the two luminaries raise will not appear distinguished, but will make a certain mixed motion. In the conjunction or op position of the luminaries their forces will be conjoined, and bring on the greatest flood and ebb. In the quadratures the sun will raise the waters which the moon depresseth. and depress the waters which the moon raiseth; and from the difference of their forces the smallest of all tides will follow. And because (as experience tells us) the force of the moon is greater than that of the sun, the greatest height of the water will happen about the third lunar hour. Out of the syzygies and quadratures the greatest tide which by the single force of the moon ought to fall out at the third lunar hour, and by the single force of the sun at the third solar hour, by the compounded forces of both must fall out in an intermediate time that approaches nearer to the third hour of the moon than to that of the sun: and, therefore, while the moon is passing from the syzygies to the quadratures, during which time the third hour of the sun precedes the third of the moon, the greatest tide will precede the third lunar hour, and that by the greatest interval a little after the octants of the moon; and by like intervals the greatest tide will follow the third lunar hour, while the moon is passing from the quadratures to the syzygies.

But the effects of the luminaries depend upon their distances from the earth; for when they are less distant their effects are greater, and when more distant their effects are less, and that in the triplicate proportion of their apparent diameters. Therefore it is that the sun in the winter time, being then in its perigee, has a greater effect, and makes the tides in the syzygies something greater, and those in the quadratures something less, *cæteris paribus*, than in the summer season; and every month the moon, while in the perigee, raiseth greater tides than at the distance of 15 days before or after, when it is in its apogee. Whence it comes to pass that two highest tides do not follow one the other in two immediately succeeding syzygies.

The effect of either luminary doth likewise depend upon its declination or distance from the equator; for if the luminary was placed at the pole, it would constantly attract all the parts of the waters, without any intension or remission of its action, and could cause no reciprocation of motion; and, therefore, as the luminaries decline from the equator towards either pole, they will by degrees lose their force, and on this account will excite lesser tides in the solstitial than in the equinoctial syzygies. But in the solstitial quadratures they will raise greater tides than in the quadratures about the equinoxes; because the effect of the moon, then situated in the equator, most exceeds the effect of the sun; therefore the greatest tides fall out in those syzygies, and the least in those quadratures, which happen about the time of both equinoxes; and the greatest tide in the syzygies is always succeeded by the least tide in the quadratures, as we find by experience. But because the sun is less distant from the earth in winter than in summer, it comes to pass that the greatest and least tides more frequently appear before than after the vernal equinox, and more frequently after than before the autumnal.

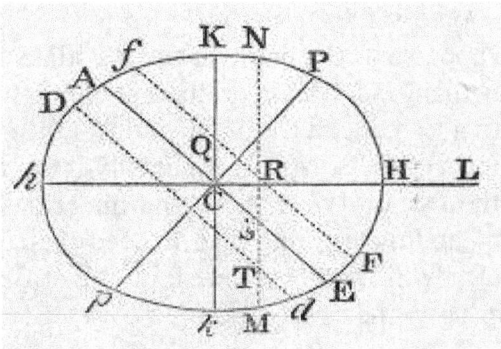

Moreover, the effects of the luminaries depend upon the latitudes of places. Let ApEP represent the earth on all sides covered with deep waters: C its centre; P, p, its poles; AE the equator: F any place without the equator: Ff the parallel of the place: Dd the correspondent parallel on the other side of the equator; L the place which the moon possessed three hours before H the place of the earth directly under it; h the opposite place; K, k, the places at 90 degrees distance; CH, Ch, the greatest heights of the sea from the centre of the earth; and CK, Ck, the least heights: and if with the axes Hh, Kk, an ellipsis is described, and by the revolution of that ellipsis about its longer axis Hh a spheroid HPK*hpk* is formed, this spheroid will nearly represent the figure of the sea; and CF, Cf, CD, Cd, will represent the sea in the places F, f, D, d. But farther: if in the said revolution of the ellipsis any point N describes the circle NM, cutting the parallels Ff, Dd in any places R, T, and the equator AE in S, CN will represent the height of the sea in all those places R, S, T, situated in this circle. Wherefore in the diurnal revolution of any place F the greatest

33

flood will be in F. at the third hour after the appulse of the moon to the meridian above the horizon; and afterwards the greatest ebb in Q, at the third hour after the setting of the moon: and then the greatest flood in f, at the third hour after the appulse of the moon to the meridian under the horizon, and, lastly, the greatest ebb in Q. at the third hour after the rising of the moon; and the latter flood in f will be less than the preceding flood in F. For the whole sea is divided into two huge and hemispherical floods, one in the hemisphere KHkC on the north side, the other in the opposite hemisphere KHkC, which we may therefore call the northern and the southern floods: these floods being always opposite the one to the other, come by turns to the meridians of all places after the interval of twelve lunar hours; and, seeing the northern countries partake more of the northern flood, and the southern countries more of the southern flood, thence arise tides alternately greater and less in all places without the equator in which the luminaries rise and set. But the greater tide will happen when the moon declines towards the vertex of the place, about the third hour after the appulse of the moon to the meridian above the horizon; and when the moon changes its declination, that which was the greater tide will be changed into a lesser; and the greatest difference of the floods will fall out about the times of the solstices, especially if the ascending node of the moon is about the first of Aries. So the morning tides in winter exceed those of the evening, and the evening tides exceed those of the morning in summer; at Plymouth by the height of one foot, but at Bristol by the height of 15 inches, according to the observations of *Colepress* and *Sturmy*.

But the motions which we have been describing suffer some alteration from that force of reciprocation which the waters [having once received] retain a little while by their *vis insita*; whence it comes to pass that the tides may continue for some time, though the actions of the luminaries should cease. This power of retaining the impressed motion lessens the difference of the alternate tides, and makes those tides which immediately succeed after the syzygies greater, and those which follow next after the quadratures less. And hence it is that the alternate tides at Plymouth and Bristol do not differ much more one from the other than by the height of a foot, or of 15 inches; and that the greatest tides of all at those ports are not the first but the third after the syzygies.

And, besides, all the motions are retarded in their passage through shallow channels, so that the greatest tides of all, in some straits and mouths of rivers, are the fourth, or even the fifth, after the syzygies.

It may also happen that the greatest tide may be the fourth or fifth after the syzygies, or fall out yet later, because the motions of the sea are retarded in passing through shallow places towards the shores: for so the tide arrives at the western coast of *Ireland* at the third lunar hour, and an hour or two after at the ports in the southern coast of the same island; as also at the islands *Cassiterides*, commonly *Sorlings*; then successively at *Falmouth*, *Plymouth*, *Portland*, the isle of *Wight*, *Winchester*, *Dover*, the mouth of the *Thames*, and *London Bridge*; spending twelve hours in this passage. But farther; the propagation of the tides may be obstructed even by the channels of the ocean itself, when they are not of depth enough, for the flood happens at the third lunar hour in the *Canary islands*; and at all those western coasts that lie towards the Atlantic ocean, as of *Ireland*, *France*, *Spain*, and all *Africa*, to the *Cape* of *Good Hope*, except in some shallow places, where it is impeded, and falls out later; and in the straits of *Gibraltar*, where, by reason of a motion propagated from the Mediterranean sea, it flows sooner. But, passing from those coasts over the breadth of the ocean to the coasts of *America*, the flood arrives first at the most eastern shores of *Brazil*, about the fourth or fifth lunar hour; then at the mouth of the river of the *Amazons* at the sixth hour, but at the neighbouring islands at the fourth hour; afterwards at the islands of *Bermudas* at the seventh hour, and at port St. *Augustin* in *Florida* at seven and a half. And therefore the tide is propagated through the ocean with a slower motion than it should be according to the course of the moon; and this retardation is very necessary, that the sea at the same time may fall between *Brazil* and *New France*, and rise at the *Canary islands*, and on the coasts of *Europe* and *Africa*, and vice versa: for the sea cannot rise in one place but by falling in another. And it is probable that the Pacific sea is agitated by the same laws: for in the coasts of *Chili* and *Peru* the highest flood is said to happen at the third lunar hour. But with what velocity it is thence propagated to the eastern coasts of *Japan*, the *Philippine* and other islands adjacent to *China*, I have not yet learned.

Farther: it may happen (p. 418) that the tide may be propagated from the ocean through different channels towards the same port, and may pass quicker through some channels than through others, in which case the same tide, divided into two or more succeeding one another, may compound new motions of different kinds. Let us suppose one tide to be divided into two equal tides, the former whereof precedes the other by the space of six hours, and happens at the third or twenty-seventh hour from the appulse of the moon to the meridian of the port. If the moon at the time of this appulse to the meridian was in the equator, every six hours alternately there would arise equal floods, which, meeting with as many equal ebbs, would so balance one the other, that, for that day, the water would stagnate, and remain quiet. If the moon then declined from the equator, the tides in the ocean would be alternately greater and less, as was said; and from hence two greater and two lesser tides would be alternately propagated towards that port. But the two greater floods would make the greatest height of the waters to fall out in the middle time betwixt both, and the Greater and lesser floods would make the waters to rise to a mean height in the middle time between them; and in the middle time between the two lesser floods the waters would rise to their least height. Thus in the space of twenty-four hours the waters would come, not twice, but once only to their greatest, and once only to their least height; and their greatest height, if the moon declined towards the elevated pole, would happen at the sixth or thirtieth hour after the appulse of the moon to the meridian and when the moon changed its declination, this flood would be changed into an ebb.

Of all which we have an example in the port of Batsham, in the kingdom of Tunquin in the latitude of 20° 50′ north. In that port, on the day which follows after the passage of the moon over the equator, the waters stagnate; when the moon declines to the north, they begin to flow and ebb, not twice, as in other ports, but once only every day; and the flood happens at the setting, and the greatest ebb at the rising of the moon. This tide increaseth with the declination of the moon till the seventh or eighth day; then for the seventh or eighth day following it decreaseth at the same rate as it had increased before, and ceaseth when the moon changeth its declination. After which the flood is immediately changed into an ebb; and thenceforth the ebb happens at the setting and the flood at the rising of the moon, till the moon again changes its declination. There are two inlets from the ocean to this port; one more direct and short between the island Hainan and the coast of Quantung, a province of China; the other round about between the same island and the coast of Cochim; and through the shorter passage the tide is sooner propagated to Batsham.

In the channels of rivers the influx and reflux depends upon the current of the rivers, which obstructs the ingress of the waters from the sea, and promotes their egress to the sea, making the ingress later and slower, and the egress sooner arid faster; and hence it is that the reflux is of longer duration that the influx, especially far up the rivers, where the force of the sea is less. So Sturmy tells us, that in the river Avon, three miles below Bristol, the water flows only five hours, but ebbs seven; and without doubt the difference is yet greater above Bristol, as at Caresham or the Bath. This difference does likewise depend upon the quantity of the flux and re flux; for the more vehement motion of the sea near the syzygies of the luminaries more easily overcoming the resistance of the rivers, will make the ingress of the water to happen sooner and to continue longer, and will therefore diminish this difference. But while the moon is approaching to the syzygies, the rivers will be more plentifully filled, their currents being obstructed by the greatness of the tides, and therefore will something more retard the reflux of the sea a little after than a little before the syzygies. Upon which account the slowest tides of all will not happen in the syzygies, but precede them a little; and I observed above that the tides before the syzygies were also retarded by the force of the sun; and from both causes conjoined the retardation of the tides will be both greater and sooner before the syzygies. All which I find to be so, by the tide-tables which Flamsted has composed from a great many observations.

By the laws we have been describing, the times of the tides are governed; but the greatness of the tides depends upon the greatness of the seas. Let C represent the centre of the earth, EADB the oval figure of the seas, CA the longer semi-axis of this oval, cB the shorter insisting at right angles upon the former, D the middle point between A and B, and ECF or eCf the angle at the centre of the earth, subtended by the breadth of the sea that terminates in the shores E, F, or e,f. Now, supposing that the point

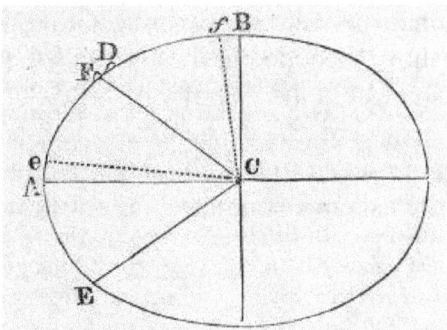

A is in the middle between the points E, F, and the point D in the middle between the points e, f, if the difference of the heights CA, CB, represent the quantity of the tide in a very deep sea surrounding the whole earth, the excess of the height CA above the height CE or CF will represent the quantity of the tide in the middle of the sea EF, terminated by the shores E, F; and the excess of the height Ce above the height Cf will nearly represent the quantity of the tide on the shores" of the same sea. Whence it appears that the tides are far less in the middle of the sea than at the shores; and that the tides at the shores are nearly as EF (p. 451, 452), the breadth of the sea not exceeding a quadrantal arc. And hence it is that near the equator, where the sea between Africa and America is narrow, the tides are far less than towards either side in the temperate zones, w r here the seas are extended wider; or on almost all the shores of the Pacific sea; as well towards America as towards China,, and within as well as without the tropics; and that in islands in the middle of the sea they scarcely rise higher than two or three feet, but on the shores of great continents are three or four times greater, and above, especially if the motions propagated from the ocean are by degrees contracted into a narrow space, and the water, to fill and empty the bays alternately, is forced to flow and ebb with great violence through shallow places; as Plymouth and Chepstow Bridge in England) at the mount of St. Michael and town of Avranches in Normandy, and at Cambaia and Peyn in the East Indies. In which places the sea, hurried in and out with great violence, sometimes lays the shores under water, sometimes leaves them dry, for many miles. Nor is the force of the influx and efflux to be broke till it has raised or depressed the water to forty or fifty feet and more. Thus also long and shallow straits that open to the sea with mouths wider and deeper than the rest of their channel (such as those about Britain and the Magellanic Straits at the eastern entry) will have a greater flood and ebb, or will more intend and remit their course, and therefore will rise higher and be depressed lower. Or the coast of South America it is said that the Pacific sea in its reflux sometimes retreats two miles, and gets out of sight of those that stand on shore. Whence in these places the floods will be also higher; but in deeper waters the velocity of influx and efflux is always less, and therefore the ascent and descent is so too. Nor in such places is the ocean known to ascend to more than six, eight, or ten feet. The quantity of the ascent I compute in the following manner

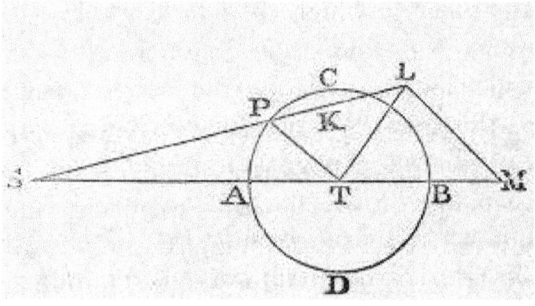

Let S represent the sun, T the earth (419, 420), P the moon, PAGB the moon's orbit. In SP take SK equal to ST and SL to SK in the duplicate ratio of SK to SP. Parallel to PT draw LM; and, supposing the mean quantity of the circum-solar force directed towards the earth to be represented by the distance ST or SK, SL will represent the quantity thereof directed towards the moon. But that force is compounded of the parts SM, LM; of which the force LM and that part of SM which is represented by TM, do disturb the motion of the moon (as appears from Prop. LXVI, and its Corollaries) In so far as the earth and moon are revolved about their common centre of gravity, the earth will be liable to the action of the like forces. But we may refer the sums as well of the forces as of the motions to the moon, and represent the sums of the forces by the lines TM and ML, which are proportional to them. The force LM, in its mean quantity, is to the force by which the moon may be revolved in an orbit, about the earth quiescent, at the distance PT in the duplicate ratio of the moon's periodic time about the earth to the earth's periodic time about the sun (by Cor. XVII, Prop. LXVI) : that is, in the duplicate ratio of 27d. 7h. 43' to 365d. 6h. 9'; or as 1000 to 178725, or 1 to $178^{29}/_{40}$. The force by which the moon may be revolved in its orb about the earth in rest, at the distance PT of 60½ semi-diameters of the earth, is to the force by which it may revolve in the same time at the distance of 60 semi-diameters as 60½ to 60; and this force is to the force of gravity with us as 1 to 60 × 60 nearly; and therefore the mean force ML is to the force of gravity at the surface of the earth as 1 × 60½ to 60 × 60 × $178^{29}/_{40}$, or 1 to 638092,6. Whence the force TM will be also given from the proportion of the lines TM, ML. And these are the forces of the sun, by which the moon's motions are disturbed.

If from the moon's orbit (p. 449) we descend to the earth's surface, those forces will be diminished in the ratio of the distances 60½ and 1; and therefore the force LM will then become 38604600 times less than the force of gravity. But this force acting equally every where upon the earth, will scarcely effect any change on the motion of the sea, and there fore may be neglected in the explication of that motion. The other force TM, in places where the sun is vertical, or in their nadir, is triple the quantity of the force ML, and therefore but 12868200 times less than the force of gravity.

Suppose now ADBE to represent the spherical surface of the earth, aDbE the surface of the water overspreading it, C the centre of both, A the place to which the sun is vertical, B the place opposite: D, E, places at 90 degrees distance from the former; ACE*mlk* a right angled cylindric canal passing through the earth's centre. The force TM in any place is as the distance of the place from the plane DE, on which a line from A

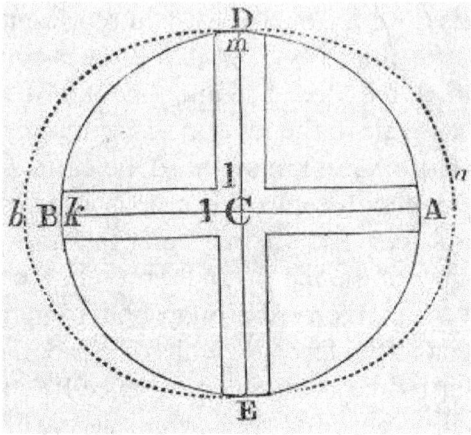

to C insists at right angles, and therefore in the part of the canal which is represented by EC*lm* is of no quantity, but in the other part AC*lk* is as the gravity at the several heights; for in descending towards the centre of the earth, gravity is (by Prop. LXXIII) every where as the height; and therefore the force TM drawing the water upwards will diminish its gravity in the leg AC*lk* of the canal in a given ratio: upon which account the water will ascend in this leg, till its defect of gravity is supplied by its greater height: nor will it rest in an equilibrium till its total gravity becomes equal to the total gravity in EC*lm*, the other leg of the canal. Because the gravity of every particle is as its distance from the earth's centre, the weight of the whole water in either leg will increase in the duplicate ratio of the height; and therefore the height of the water in the leg AC*lk* will be to the height thereof in the leg C*lm*E in the subduplicate ratio of the number 12868201 to 12808200, or in the ratio of the number 25623053 to the number 25623052, and the height of the water in the leg EC*lm* to the difference of the heights, as 25623052 to 1. But the height in the leg EC*lm* is of 19615800 Paris feet, as has been lately found by the mensuration of the French; and, therefore, by the preceding analogy, the difference of the heights comes out $9\frac{1}{5}$ inches of the Paris foot; and the sun's force will make the height of the sea at A to exceed the height of the same at E by 9 inches. And though the water of the canal ACE*mlk* be supposed to be frozen into a hard and solid consistence, yet the heights thereof at A and E, and all other intermediate places, would still remain the same.

Let Aa (in the following figure) represent that excess of height of 9 inches at A, and hf the excess of height at any other place h; and upon DC let fall the perpendicular G, meeting the globe of the earth in F: and because the distance of the sun ib so great that all the right lines drawn thereto may be considered as parallel, the force TM in any place will be to the same force in the place A as the sine FG to the radius AC. And, therefore, since those forces tend to the sun in the direction of

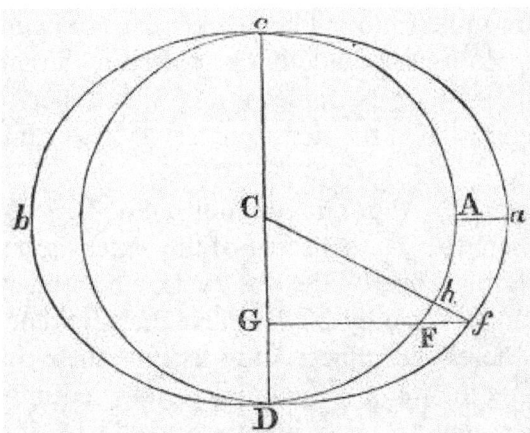

parallel lines, they will generate the parallel heights F An, in the same ratio; and there fore the figure of the water Dfaeb will be a spheroid made by the revolution of an ellipsis about its longer axis ab. And the perpendicular height fh will be to the oblique height F as G to C, or as FG to AC: and there fore the height fh is to the height Art in the duplicate ratio of FG to AC, that is, in the ratio of the versed sine of double the angle DC to double the radius, and is thence given. And hence to the several moments of the apparent revolution of the sun about the earth we may infer the proportion of the ascent and descent of the waters at any given place under the equator, as well as of the diminution of that ascent and descent, whether arising from the latitude of places or from the sun s declination; viz., that on account of the latitude of places, the ascent and descent of the sea is in all places diminished in the duplicate ratio of the cosines of latitude; and on account of the sun's declination, the ascent and descent under the equator is diminished in the duplicate ratio of the cosine of declination. And in places without the equator the half sum of the morning and evening ascents (that is, the mean ascent) is diminished nearly in the same ratio.

Let S and L respectively represent the forces of the sun and moon placed in the equator, and at their mean distances from the earth; R the radius; T and V the versed sines of double the complements of the sun and moon's declinations to any given time; D and E the mean apparent diameters of the sun and moon: and, supposing F and G to be their apparent diameters to that given time, their forces to raise the tides under the equator will be, in the syzygies VG^3

$2RE^3 \, L + TF^3$

$2RD^3 \, S$; in the quadratures, VG^3

$2RE^3 \, L - TF^3$

$2RD^3 \, S$. And if the same ratio is likewise observed under the parallels, from observations accurately made in our northern climates we may determine the proportion of the forces L and S; and then by means of this rule predict the quantities of the tides to every syzygy and quadrature.

At the mouth of the river Avon, three miles below Bristol (p. 450 to 453), in spring and autumn, the whole ascent of the water in the conjunction or opposition of the luminaries (by the observation of Sturmy) is about 45 feet, but in the quadratures only 25. Because the apparent diameters of the luminaries are not here determined, let us assume them in their mean quantities, as well as the moon's declination in the equinoctial quadratures in its mean quantity, that is, 23½°; and the versed sine of double its complement will be 1682, supposing the radius to be 1000. But the declinations of the sun in the equinoxes and of the moon in the syzygies are of no quantity, and the versed sines of double the complements are each 2000. Whence those forces become L + S in the syzygies, and 1682 $2000L - S$ in the quadrature; respectively proportional to the heights of the tides of 45 and 25 feet, or of 9 and 5 paces. And, therefore, multiplying the extremes and the means, we have 5L + 5S = 15138

$2000L - 9S$ or L = 28000

$5138S = 5^5/_{11}S$.

But farther; I remember to have been told that in summer the ascent of the sea in the syzygies is to the ascent thereof in the quadratures as about 5 to 4. In the solstices themselves it is probable that the proportion may be something less, as about 6 to 5; whence it would follow that L is = $5^1/_6$S [for then the proportion is 1682

$2000L + 1682$

$2000S : L - 1682$

$2000S :: 6 : 5$]. Till we can more certainly determine the proportion from observation, let us assume L = $5^1/_3$S; and since the heights of the tides are as the forces which excite them, and the force of the sun is able to raise the tides to the height of nine inches, the moon's force will be sufficient to raise the same to the height of four feet. And if we allow that this height may be doubled, or perhaps tripled, by that force of reciprocation which we observe in the motion of the waters, and by which their motion once begun is kept up for some time, there will be force enough to generate all that quantity of tides which we really find in the ocean.

Thus we have seen that these forces are sufficient to move the sea. But. so far as I can observe, they will not be able to produce any other effect sensible on our earth; for since the weight of one grain in 4000 is not sensible in the nicest balance: and the sun's force to move the tides is 12868200 less than the force of gravity; arid the sum of the forces of both moon and sun, exceeding the sun's force only in the ratio of $6\frac{1}{3}$ to 1, is still 2032890 times less than the force of gravity; it is evident that both forces together are 500 times less than what is required sensibly to increase or diminish the weight of any body in a balance. And, therefore, they will not sensibly move any suspended body; nor will they produce any sensible effect on pendulums, barometers, bodies swimming in stagnant water, or in the like statical experiments. In the atmosphere, indeed, they will excite such a flux and reflux as they do in the sea, but with so small a motion that no sensible wind will be thence produced.

If the effects of both moon and sun in raising the tides (p. 454), as well as their apparent diameters, were equal among themselves, their absolute forces would (by Cor. XIV, Prop. LXVI) be as their magnitudes. But the effect of the moon is to the effect of the sun as about $5\frac{1}{3}$ to 1; and the moon's diameter less than the sun's in the ratio of $31\frac{1}{2}$ to $32\frac{1}{3}$, or of 45 to 46. Now the force of the moon is to be increased in the ratio of the effect directly, and in the triplicate ratio of the diameter inversely. Whence the force of the moon compared with its magnitude will be to the force of the sun compared with its magnitude in the ratio compounded of $5\frac{1}{3}$ to 1, and the triplicate of 45 to 46 inversely, that is, in the ratio of about $5^{7}/_{10}$ to 1. And therefore the moon, in respect of the magnitude of its body, has an absolute centripetal force greater than the sun in respect of the magnitude of its body in the ratio to $5^{7}/_{10}$ to 1, and is therefore more dense in the same ratio.

In the time of 27d.7h.43', in which the moon makes its revolution about the earth, a planet may be revolved about the sun at the distance of 18.95 1 diameters of the sun from the sun's centre, supposing the mean apparent diameter of the sun to be 32 1/5'; and in the same time the moon may be revolved about the earth at rest, at the distance of 30 of the earth's diameters. If in both cases the number of diameters was the same, the absolute circum-terrestrial force would (by Cor. II, Prop. LXXII) be to the absolute circum-solar force as the magnitude of the earth to the magnitude of the sun. Because the number of the earth's diameters is greater in the ratio of 30 to 18,954, the body of the earth will be less in the triplicate of that ratio, that is, in the ratio of $3^{28}/_{29}$ to 1. Wherefore the earth's force, for the magnitude of its body, is to the sun's force, for the magnitude of its body, as $3^{28}/_{29}$ to 1: and consequently the earth's density to the sun's will be in the same ratio. Since, then, the moon's density is to the sun's density as $5^{7}/_{10}$ to 1, the moon's density will be to the earth's density as $5^{7}/_{10}$ to $3^{28}/_{29}$, or as 23 to 16. Wherefore since the moon's magnitude is to the earth s magnitude as about 1 to 41½, the moon's absolute centripetal force will be to the earth's absolute centripetal force as about 1 to 29, and the quantity of matter in the moon to the quantity of matter in the earth in the same-ratio. And hence the common centre of gravity of the earth and moon is more exactly determined than hitherto has been done; from the knowledge of which we may now infer the moon's distance from the earth with greater accuracy. But I would rather wait till the proportion of the bodies of the moon and earth one to the other is more exactly defined from the phaenomena of the tides, hoping that in the mean time the circumference of the earth may be measured from more distant stations than any body has yet employed for this purpose.

Thus I have given an account of the system of the planets. As to the fixed stars, the smallness of their annual parallax proves them to be re moved to immense distances from the system of the planets: that this parallax is less than one minute is most certain; and from thence it follows that the distance of the fixed stars is above 360 times greater than the distance of Saturn from;he sun. Such as reckon the earth one of the planets, and the sun one of the fixed stars, may remove the fixed stars to yet greater distances by the following arguments: from the annual motion of the earth there would happen an apparent transposition of the fixed stars, one in respect of another, almost equal to their double parallax: but the greater and nearer stars, in respect of the more remote, which are only seen by the telescope, have not hitherto been observed to have the least motion. If we should suppose that motion to be but less than 20″, the distance of the nearer fixed stars would exceed the mean distance of Saturn by above 2000 times. Again: the disk of Saturn, which is only 17″ or 18″ in diameter, receives but about 1/2100000000 of the sun's light; for so much less is that disk than the whole spherical surface of the orb of Saturn. Now if we suppose Saturn to reflect about 1/4 of this light, the whole light reflected from its illuminated hemisphere will be about 1/4200000000 of the whole light emitted from the sun's hemisphere: and, therefore, since light is rarefied in the duplicate ratio of the distance from the luminous body, if the sun was $10000 \sqrt{42}$ times more distant than Saturn, it would yet ap pear as lucid as Saturn now does without its ring, that is, something more lucid than a fixed star of the first magnitude. Let us, therefore, suppose that the distance from which the sun would shine as a fixed star exceeds that of Saturn by about 100,000 times, and its apparent diameter will be $7^{v}.16^{vi}$ and its parallax arising from the annual motion of the earth 13′′′′: and so great will be the distance, the apparent diameter, and the parallax of the fixed stars of the first magnitude, in bulk and light equal to our sun.

Some may, perhaps, imagine that a great part of the light of the fixed stars is intercepted and lost in its passage through so vast spaces, and upon that account pretend to place the fixed stars at nearer distances; but at this rate the remoter stars could be scarcely seen. Suppose, for example, that of the light perish in its passage from the nearest fixed stars to us; then | will twice perish in its passage through a double space, thrice through a triple, and so forth. And, therefore, the fixed stars that are at a double distance will be 16 times more obscure, viz., 4 times more obscure on account of the diminished apparent diameter; and, again, 4 times more on account of the lost light. And, by the same argument, the fixed stars at a triple distance will be $9 \times 4. \times 4$, or 144 times more obscure; and those at a quadruple distance will be $16 \times 4 \times 4 \times 4$, or 1024 times more obscure: but so great a diminution of light is no ways consistent with the phenomena and with that hypothesis which places the fixed stars at different distances.

The fixed stars being, therefore, at such vast distances from one another (p. 460, 461), can neither attract each other sensibly, nor be attracted by our sun. But the comets must unavoidably be acted on by the circum-solar force; for as the comets were placed by astronomers above the moon, because they were found to have no diurnal parallax, so their annual parallax is a convincing proof of their descending into the regions of the planets. For all the comets which move in a direct course, according to the order of the signs, about the end of their appearance become more than ordinarily slow, or retrograde, if the earth is between them and the sun; and more than ordinarily swift if the earth is approaching to a heliocentric opposition with them. Whereas, on the other hand, those which move against the order of the signs, towards the end of their appearance, appear swifter than they ought to be if the earth is between them and the sun; and slower, and perhaps retrograde, if the earth is in the other side of its orbit. This is occasioned by the motion of the earth in different situations. If the earth go the same way with the comet, with a swifter motion, the comet becomes retrograde; if with a slower motion, the comet becomes slower, however; and if the earth move the contrary way, it be comes swifter; and by collecting the differences between the slower and swifter motions, and the sums of the more swift and retrograde motions, and comparing them with the situation and motion of the earth from, whence they arise, I found, by means of this parallax, that the distances of the comets at the time they cease to be visible to the naked eye are always less than the distance of Saturn, and generally even less than the distance of Jupiter.

The same thing may be collected from the curvature of the way of the comets (p. 462). These bodies go on nearly in great circles while their motion continues swift; but about the end of their course, when that part of their apparent motion which arises from the parallax bears a greater proportion to their whole apparent motion, they commonly deviate from those circles; and when the earth goes to one side, they deviate to the other; and this deflection, because of its corresponding with the motion of the earth, must arise chiefly from the parallax; and the quantity there of is so considerable, as, by my computation, to place the disappearing comets a good deal lower than Jupiter. Whence it follows, that, when they approach nearer to us in their perigees and perihelions, they often descend below the orbits of Mars and the inferior planets.

Moreover, this nearness of the comets is confirmed by the annual parallax of the orbit, in so far as the same is— pretty nearly collected by the supposition that the comets move uniformly in right lines. The method of collecting the distance of a comet according to this hypothesis from four observations (first attempted by Kepler, and perfected by Dr. Wallis and Sir Christopher Wren) is well known and the comets reduced to this regularity generally pass through the middle of the planetary region. So the comets of the year 1607 and 1618, as their motions are defined by Kepler, passed between the sun and the earth: that of the year 16?4 below the orbit of Mars; and that in 1680 below the orbit of Mercury, as its motion was defined by Sir Christopher Wren and others. By a like rectilinear hypothesis, Hevelius placed all the comets about which we have any observations below the orbit of Jupiter. It is a false notion, there fore, and contrary to astronomical calculation, which some have entertained, who, from the regular motion of the comets, either remove them into the regions of the fixed stars, or deny the motion of the earth:

where as their motions cannot be reduced to perfect regularity, unless we suppose them to pass through the regions near the earth in motion; and these are the arguments drawn from the parallax, so far as it can be determined without an exact knowledge of the orbits and motions of the comets.

The near approach of the comets is farther confirmed from the light of their heads (p. 463, 465); for the light of a celestial body, illuminated by the sun, and receding to remote parts, is diminished in the quadruplicate proportion of the distance; to wit, in one duplicate proportion on account of the increase of the distance from the sun; and in another duplicate proportion on account of the decrease of the apparent diameter. Hence it may be inferred, that Saturn being at a double distance, and having its apparent diameter nearly half of that of Jupiter, must appear about 10 times more obscure; and that, if its distance were 4 times greater, its light would be 256 times less; and therefore would be hardly perceivable to the naked eye. But now the comets often equal Saturn's light, without exceeding him in their apparent diameters. So the comet of the year 1668, according to Dr. Hooke's observations, equalled in brightness the light of a fixed star of the first magnitude; and its head, or the star in the middle of the coma, appeared, through a telescope of 15 feet, as lucid as Saturn near the horizon; but the diameter of the head was only 25″ that is, almost the same with the diameter of a circle equal to Saturn and his ring. The coma or hair surrounding the head was about ten times as broad; namely, 4 1/6 min. Again; the least diameter of the hair of the comet of the year 1682, observed by Mr. Flamsted with a tube of 16 feet and measured with the micrometer, was 2′ 0″; but the nucleus, or star in the middle, scarcely possessed the tenth part of this breadth, and was therefore only 11″ or 12″ broad; but the light and clearness of its head exceeded that of the year 1680, and was equal to that of the stars of the first or second magnitude. Moreover, the comet of the year 1665, in April, as Hevelius informs us, exceeded almost all the fixed stars in splendor, and even Saturn itself, as being of a much more vivid colour; for this comet was more lucid than that which appeared at the end of the foregoing year and was compared to the stars of the first magnitude. The diameter of the coma was about 6′; but the nucleus, compared with the planets by means of a telescope, was plainly less than Jupiter, and was sometimes thought less, sometimes equal to the body of Saturn within the ring. To this breadth add that of the ring, and the whole face of Saturn will be twice as great as that of the comet, with a light not at all more intense; and therefore the comet was nearer to the sun than Saturn. From the proportion of the nucleus to the whole head found by these observations, and from its breadth, which seldom exceeds 8′ or 12′, it appears that the stars of the comets are most commonly of the same apparent magnitude as the planets; but that their light may be compared oftentimes with that of Saturn, and sometimes exceeds it. And hence it is certain that in their perihelia their distances can scarcely be greater than that of Saturn. At twice that distance, the light would be four times less, which besides by its dim paleness would be as much inferior to the light of Saturn as the light of Saturn is to the splendor of Jupiter: but this difference would be easily observed. At a distance ten times greater, their bodies must be greater than that of the sun; but their light would be 100 times fainter than that of Saturn. And at distances still greater, their bodies would far exceed the sun; but, being in such dark regions, they must be no longer visible. So impossible is it to place the comets in the middle regions between the sun and fixed stars, accounting the sun as one of the fixed stars: for certainly they would receive no more light there from the sun than we do from the greatest of the fixed stars.

So far we have gone without considering that obscuration which comets suffer from that plenty of thick smoke which encompasseth their heads, and through which the heads always shew dull as through a cloud; for by how much the more a body is obscured by this smoke, by so much th.2 more near it must be allowed to come to the sun, that it may vie with the planets in the quantity of light which it reflects: whence it is probable that the comets descend far below the orbit of Saturn, as we proved before from their parallax. But, above all, the thing is evinced from their tails, which must be owing either to the sun's light reflected from a smoke arising from them, and dispersing itself through the aether, or to the light of their own heads.

In the former case we must shorten the distance of the comets, lest we be obliged to allow that the smoke arising from their heads is propagated through such a vast extent of space, and with such a velocity of expansion, as will seem altogether incredible; in the latter case the whole light of both head and tail must be ascribed to the central nucleus. But, then, if we suppose all this light to be united and condensed within the disk of the nucleus, certainly the nucleus will by far exceed Jupiter itself in splendor, especially when it emits a very large and lucid tail. If, therefore, under a less apparent diameter, it reflects more light, it must be much more illuminated by the sun, and therefore much nearer to it. So the comet that appeared Dec. 12 and 15, O.S. Anno 1679, at the time it emitted a very shining tail, whose splendor was equal to that of many stars like Jupiter, if their light were dilated and spread through so great a space, was, as to the magnitude of its nucleus, less than Jupiter (as Mr. Flamsted observed), and therefore was much nearer to the sun: nay, it was even less than Mercury. For on the 17th of that month, when it was nearer to the earth, it appeared to Cassini through a telescope of 35 feet a little less than the globe of Saturn. On the 8th of this month, in the morning, Dr. Halley saw the tail, appearing broad and very short, and as if it rose from the body of the sun itself, at that time very near its rising. Its form was like that of an extraordinary bright cloud ; nor did it disappear till the sun itself began to be seen above the horizon. Its splendor, therefore, exceeded the light of the clouds till the sun rose, and far surpassed that of all the stars together, as yielding only to the immediate brightness of the sun itself. Neither Mercury, nor Venus, nor the moon itself, are seen so near the rising sun. Imagine all this dilated light collected together, and to be crowded into the orbit of the comet's nucleus which was less than Mercury; by its splendor, thus increased, becoming so much more conspicuous, it will vastly exceed Mercury, and therefore must be nearer to the sun. On the 12th and 15th of the same month, this tail, extending itself over a much greater space, appeared more rare; but its light was still so vigorous as to become visible when the fixed stars were hardly to be seen, and soon after to appear like a fiery beam shining in a wonderful manner. From its length, which was 40 or 50 degrees, and its breadth of 2 degrees, we may compute what the light of the whole must be.

This near approach of the comets to the sun is confirmed from the situation they are seen in when their tails appear most resplendent; for when the head passes by the sun, and lies hid under the solar rays, very bright and shining ta Is, like fiery beams, are said to issue from the horizon; but afterwards, when the head begins to appear, and is got farther from the sun, that splendor always decreases, and turns by degrees into a paleness like to that of the milky way, but much more sensible at first; after that vanishing gradually. Such was that most resplendent comet described by Aristotle, Lib. 1, Meteor. 6. "The head thereof could not be seen, because it set before the sun, or at least was hid under the sun's rays; but the next day it was seen as well as might be; for, having left the sun but a very little way, it set immediately after it; and the scattered light of the head obscured by the too great splendour (of the tail) did not yet appear. But afterwards (says Aristotle), when the splendour of the tail was now diminished (the head of), the comet recovered its native brightness. And the splendour of its tail reached now to a third part of the heavens (that is, to 60°). It appeared in the winter season, and, rising to Orion's girdle, there vanished away." Two comets of the same kind are described by Justin, Lib. 37, which, according to his account, "shined so bright, that the whole heaven seemed to be on fire; and by their greatness filled up a fourth part of the heavens, and by their splendour exceeded that of the sun." By which last words a near position of these bright comets and the rising or setting sun is intimated (p. 494, 495). We may add to these the comet of the year 1101 or 1106, "the star of which was small and obscure (like that of 1680); but the splendour arising from it extremely bright, reaching like a fiery beam to the east and north," as Hevelius has it from Simeon, the monk of Durham. It appeared at the beginning of February about the evening in the south-west. From this and from the situation of the tail we may infer that the head was near the sun. Matthew Paris says, "it was about one cubit from the sun; from the third [or rather the sixth] to the ninth hour sending out a long stream of light." The comet of 1264, in July, or about the solstice, preceded the rising sun, sending out its beams with a great light towards the west as far as the middle of the heavens; and at the beginning it ascended a little above the horizon: but as the sun went forwards it retired every day farther from the horizon, till it passed by the very middle of the heavens. It is said to have been at the beginning large and bright, having a large coma, which decayed from day to day. It is described in *Append. Matth. Paris, Hist. Ang.* after this manner: "An. Christi 1265, there appeared a comet so wonderful, that none then living had ever seen the like; for, rising from the east with a great brightness, it extended itself with a great light as far as the middle of the hemisphere towards the west." The Latin original being somewhat barbarous and obscure, it is here subjoined. *Ab oriente enim cum magno fulgore surg??s, usque ad medium hemisphaerii versus occidentem, omnia per lucid pertrahebai.*

"In the year 1401 or 1402, the sun being got below the horizon, there appeared in the west a bright and shining comet, sending out a tail up wards, in splendor like a flame of fire, and in form like a spear, darting its rays from west to east. When the sun was sunk below the horizon, by the lustre of its own rays it enlightened all the borders of the earth, not permitting the other stars to shew their light, or the shades of night to darken the air, because its light exceeded that of the others, and extended itself to the upper part of the heavens, flaming," &c., *Hist. Byzant. Duc. Mich. Nepot.* From the situation of the tail of this comet, and the time of its first appearance, we may infer that the head was then near the sun, and went farther from him every day; for that comet continued three months. In the year 1527, Aug. 11, about four in the morning, there was seen almost throughout Europe a terrible comet in Leo, which continued flaming an hour and a quarter every day. It rose from the east, and ascended to the south and west to a prodigious length. It was most conspicuous to the north, and its cloud (that is, its tail) was very terrible; having, according to the fancies of the vulgar, the form of an arm a little bent holding a sword of a vast magnitude. In the year 1618, in the end of November, there began a rumour, that there appeared about sun-rising a bright beam, which was the tail of a comet whose head was yet concealed within the brightness of the solar rays. On Nov. 24, and from that time, the comet itself appeared with a bright light, its head and tail being extremely resplendent. The length of the tail, which was at first 20 or 30 deg., increased till December 9, when it arose to 75 deg,, but with a light much more faint and dilute than at the beginning. In the year 1668, March 5, N.S., about 7 in the evening, *P. Valent. Estancius*, being in Brazil, saw a comet near the horizon in the south-west. Its head was small, and scarcely discernible, but its tail extremely bright and refulgent, so that the reflection of it from the sea was easily seen by those who stood upon the shore. This great splendor lasted but three days, decreasing very remarkably from that time. The tail at the beginning extended itself from west to south, and in a situation almost parallel to the horizon, appearing like a shining beam 23 deg. in length. Afterwards, the light decreasing, its magnitude increased till the comet ceased to be visible; so that Cassini, at Bologna, saw it (Mar. 10, 11, 12) rising from the horizon 32 deg. in length. In Portugal it is said to have taken up a fourth part of the heavens (that is, 45 deg.), extending itself from west to east with a notable brightness; though the whole of it was not seen, because the head in this part of the world always lay hid below the horizon. From the increase of the tail it is plain that the head receded from the sun. and was nearest to it at the beginning, when the tail appeared brightest.

To all these we may add the comet of 1680, whose wonderful splendor at the conjunction of the head with the sun was above described. But so great a splendor argues the comets of this kind to have really passed near the fountain of light, especially since the tails never shine so much in their opposition to the sun; nor do we read that fiery beams have ever appeared there.

Lastly, the same thing is inferred (p. 466; 407) from the light of the heads increasing in the recess of the comets from the earth towards the sun, and decreasing in their return from the sun towards the earth; for so the last comet of the year 1665 (by the observation of Hevelius), from the time that it was first seen, was always losing of its apparent motion, and therefore had already passed its perigee: yet the splendor of its head was daily increasing, till, being hid by the sun's rays, the comet ceased to appear. The comet of the year 1683 (by the observation of the same Hevelius), about the end of July, when it first appeared, moved at a very slow rate, advancing only about 40 or 45 minutes in its orbit in a day's time. But from that time its diurnal motion was continually upon the increase till September 4, when it arose to about 5 degrees; and therefore in all this interval of time the comet was approaching to the earth. Which is likewise proved from the diameter of its head measured with a micrometer; for, August the 6th, Hevelius found it only 6′ 5″, including the coma; which, September 2, he observed 9′ 7″. And therefore its head appeared far less about the beginning than towards the end of its motion, though about the beginning, because nearer to the sun, it appeared far more lucid than towards the end, as the same Hevelius declares. Wherefore in all this interval of time, on account of its recess from the sun, it decreased in splendor, notwithstanding its access towards the earth. The comet of the year 1618, about the middle of December, and that of the year 1680, about the end of the same month, did both move with their greatest velocity, and were therefore then in their perigees: but the greatest splendor of their heads was seen two weeks before, when they had just got clear of the sun's rays: and the greatest splendor of their tails a little more early, when yet nearer to the sun. The head of the former comet, according to the observations of Cysatus, Dec. 1, appeared greater than the stars of the first magnitude: and, Dec. 16 (being then in its perigee), of a small magnitude, and the splendor or clearness was much diminished. Jan. 7, Kepler, being uncertain about the head, left off observing. Dec. 12, the head of the last comet was seen and observed by Flamsted at the distance of 9 degrees from the sun, which a star of the third magnitude could hardly have been. December 15 and 17, the same appeared like a star of the third magnitude, its splendor being diminished by the bright clouds near the setting sun. Dec. 26, when it moved with the greatest swiftness, and was almost in its perigee, it was inferior to *Os Pegasi*, a star of the third magnitude. Jan. 3, it appeared like a star of the fourth; Jan. 9, like a star of the fifth. Jan. 13. it disappeared, by reason of the brightness of the moon, which was then in its increase. Jan. 25, it was scarcely equal to the stars of the seventh magnitude. If we take equal times on each hand of the perigee, the heads placed at remote distances would have shined equally before and after, because of their equal distances from the earth. That in one case they shined very bright, and in the other vanished, is to be ascribed to the nearness of the sun in the first case, and his distance in the other; and from the great difference of the light in these two cases we infer its great nearness in the first of them: for the light of the comets uses to be regular, and to appear greatest when their heads move the swiftest, and are therefore in their perigees; excepting in so far as it is increased by their nearness to the sun.

From thee things I at last discovered why the comets frequent so much the region of the sun. If they were to be seen in the regions a great way beyond Saturn, they must appear oftener in these parts of the heavens that are opposite to the sun; for those which are in that situation would be nearer to the earth, and the interposition of the sun would obscure the others: but, looking over the history of comets, I find that four or five times more have been seen in the hemisphere towards the sun than in th-3 opposite hemisphere; besides, without doubt, not a few which have been hid by the light of the sun; for comets descending into our parts neither emit tails, nor are so well illuminated by the sun, as to discover themselves to our naked eyes, till they are come nearer to us than Jupiter. But the far greater part of that spherical space, which is described about the sun with so small an interval, lies on that side of the earth which regards the sun, and the comets in that greater part are more strongly illuminated, as being for the most part nearer to the sun: besides, from the remarkable eccentricity of their orbits, it comes to pass that their lower apsides are much nearer to the sun than if their revolutions were performed in circles concentric to the sun.

Hence also we understand why the tails of the comets, while their heads are descending towards the sun, always appear short and rare, and are seldom said to have exceeded 15 or 20 deg. in length; but in the recess of the heads from the sun often shine like fiery beams, and soon after reach to 40, 50, 60, 70 deg. in length, or more. This great splendor and length of the tails arises from the heat which the sun communicates to the comet as it passes near it. And thence, I think, it may be concluded, that all the comets that have had such tails have passed very near the sun.

Hence also we may collect that the tails arise from the atmospheres of the heads (p. 487 to 488): but we have had three several opinions about the tails of comets; for some will have it that they are nothing else but the beams of the sun's light transmitted through the comets heads, which they suppose to be transparent; others, that they proceed from the refraction which light suffers in passing from the comet's head to the earth; and, lastly, others, that they are a sort of clouds or vapour constantly rising from the cornets heads, and tending towards the parts opposite to the sun. The first is the opinion of such as are yet unacquainted with optics; for the beams of the sun are not seen in a darkened room, but in consequence of the light that is reflected from them by the little particles of dust and smoke which are always flying about in the air; and hence it is that in air impregnated with thick smoke they appear with greater brightness, and are more faintly and more difficultly seen in a finer air; but in the heavens, where there is no matter to reflect the light, they are not to be seen at all. Light is not seen as it is in the beams, but as it is thence reflected to our eyes; for vision is not made but by rays falling upon the eyes, and therefore there must be some reflecting matter in those parts where the tails of comets are seen; and so the argument turns upon the third opinion; for that reflecting matter can be no where found but in the place of the tail, because otherwise, since all the celestial spaces are equally illuminated by the sun's light, no part of the heavens could appear with more splendor than another. The second opinion is liable to many difficulties. The tails of comets are never seen variegated with those colours which ever use to be inseparable from refraction; and the distinct transmission of the light of the fixed stars and planets to us is a demonstration that the aether or celestial medium is not endowed with any refractive power. For as to what is alledged that the fixed stars have been sometimes seen by the Egyptians environed with a coma or capillitium because that has but rarely happened, it is rather to be ascribed to a casual refraction of clouds, as well as the radiation and scintillation of the fixed stars to the refractions both of the eyes and air; for upon applying a telescope to the eye, those radiations and scintillations immediately disappear. By the tremulous agitation of the air and ascending vapours, it happens that the rays of light are alternately turned aside from the narrow space of the pupil of the eye; but no such thing can have place in the much wider aperture of the object-glass of a telescope; and hence it is that a scintillation is occasioned in the former case which ceases in the latter; and this cessation in the latter case is a demonstration of the regular trans mission of light through the heavens without any sensible refraction. But, to obviate an objection that may be made from the appearing of no tail in such comets as shine but with a faint light, as if the secondary rays were then too weak to affect the eyes, and for this reason it is that the tails of the fixed stars do not appear, we are to consider that by the means of telescopes the light of the fixed stars may be augmented above an hundred fold and yet no tails are seen; that the light of the planets is yet more copious without any tail, but that comets are seen sometimes with huge tails when the light of their heads is but faint and dull; for so it happened in the comet of the year 1680, when in the month of December it was scarcely equal in light to the stars of the second magnitude and yet emitted a notable tail, extending to the length of 40, 50, 60. or 70, and upwards; and afterwards, on the 27th and 28th of January, the head appeared but as a star of the seventh magnitude; but the tail (as was said above), with a light that was sensible enough, though faint, was stretched out to 6 or 7 degrees in length, and with a languishing light that was more difficultly seen, even to 12 and upwards. But on the 9th and 10th of February, when to the naked eye the head appeared no more, I saw through a telescope the tail of 2 in length. But farther: if the tail was owing to the refraction of the celestial matter, and did deviate from the opposition of the sun, according as the figure of the heavens requires, that deviation, in the same places of the heavens, should be always directed towards the same parts: but the comet of the year 1680, December 28d. 8½h. P.M. at London, was seen in Pisces, 8° '41, with latitude north 28° 6', while the sun was in Capricorn 18° 26'. And the comet of the year 1577, December 29, was in Pisces 8° 41', with latitude north 28° 40'; and the sun, as before, in about Capricorn 18° 26'. In both cases the situation of the earth was the same, and the comet appeared in the same place of the heavens; yet in the former case the tail of the comet (as well by my observations as by the observations of others) deviated from the opposition of the sun towards the north by an angle of 4½ degrees, whereas in the latter there was (according to the observation of Tycho a deviation of 21 degrees towards the south. The refraction, therefore, of the heavens being thus disproved, it remains that the phaenomena of the tails of comets must be derived from some reflecting matter. That vapours sufficient to fill such immense spaces may arise from the comet's atmospheres, may be easily understood from what follows.

It is well known that the air near the surface of our earth possesses a space about 1200 times greater than water of the same weight; and there fore a cylindric column of air 1200 feet high is of equal weight with a cylinder of water of the same breadth, and but one foot high. But a cylinder of air reaching to the top of the atmosphere is of equal weight with a cylinder of water about 33 feet high; and therefore if from the whole cylinder of air the lower part of 1200 feet high is taken away, the remaining upper part will be of equal weight with a cylinder of water 32 feet high. Wherefore at the height of 1200 feet, or two furlongs, the weight of the incumbent air is less, and consequently the rarity of the compressed air greater, than near the surface of the earth in the ratio of 33 to 32. And, having this ratio, we may compute the rarity of the air in all places whatsoever (by the help of Cor. Prop. XXII, Book II), supposing the expansion thereof to be reciprocally proportional to its compression; and this proportion has been proved by the experiments of Hooke and others. The result of the computation I have set down in the following table, in the first column of which you have the height of the air in miles, whereof 4000 m:ike a semi-diameter of the earth; in the second the compression of the air, or the incumbent weight; in the third its rarity or expansion, supposing gravity to decrease in the duplicate ratio of the distances from the earth's centre. And the Latin numeral characters are here used for certain numbers of ciphers, as 0,xvii 1224 for 0,00000000000000001224, and 26950 xv for 26956000000000000000.

	AIR's	
Height.	Compression.	Expansion.
0	33	1
5	17,8515	1.8486
10	9,6717	3.4151
20	2.852	11.571
40	0,2525	136.83
400	O.xvii 1224	26936 xv
4000	O.cv 4465	73907 cii
40000	O.cxcii 1628	20263 clxxxix
400000	0,ccx 7895	41798 ccvii
4000000	0,ccxii 9878	33414 ccix
Infinite.	O.ccxii 6041	54622 ccix

But from this table it appears that the air, in proceeding upwards, is rarefied in such manner, that a sphere of that air which is nearest to the earth, of but one inch in diameter, if dilated with that rarefaction which it would have at the height of one semi-diameter of the earth, would fill all the planetary regions as far as the sphere of Saturn, and a great way beyond; and at the height of ten semi-diameters of the earth would fill up more space than is contained in the whole heavens on this side the fixed stars, according to the preceding computation of their distance. And though, by reason of the far greater thickness of the atmospheres of comets, and the great quantity of the circum-solar centripetal force, it may happen that the air in the celestial spaces, and in the tails of comets, is not so vastly rarefied, yet from this computation it is plain that a very small quantity of air and vapour is abundantly sufficient to produce all the appearances of the tails of comets; for that they are indeed of a very notable rarity appears from the shining of the stars through them. The atmosphere of the earth, illuminated by the sun's light, though but of a few miles in thickness, obscures arid extinguishes the light not only of all the stars, but even of the moon itself; whereas the smallest stars are seen to shine through the immense thickness of the tails of comets, likewise illuminated by the sun, without the least diminution of their splendor.

Kepler ascribes the ascent of the tails of comets to the atmospheres of their heads, and their direction towards the parts opposite to the sun to the action of the rays of light carrying along with them the matter of the comets tails; and without any great incongruity we may suppose that, in so free spaces, so fine a matter as that of the aether may yield to the action of the rays of the sun's light, though those rays are not able sensibly to move the gross substances in our parts, which are clogged with so palpable a resistance. Another author thinks that there may be a sort of particles of matter endowed with a principle of levity as well as others are with a power of gravity; that the matter of the tails of comets may be of the former sort, and that its ascent from the sun may be owing to its levity; but, considering the gravity of terrestrial bodies is as the matter of the bodies, and therefore can be neither more nor less in the same quantity of matter, I am inclined to believe that this ascent may rather proceed from the rarefaction of the matter of the comets tails. The ascent of smoke in a chimney is owing to the impulse of the air with which it is entangled. The air rarefied by heat ascends, because its specific gravity is diminished, and in its ascent carries along with it the smoke with which it is engaged. And why may not the tail of a comet rise from the sun after the same manner? for the sun's rays do not act any way upon the mediums which they pervade but by reflection and refraction; and those reflecting particles heated by this action, heat the matter of the aether which is involved with them. That matter is rarefied by the heat which it acquires, and because by this rarefaction the specific gravity, with which it tended towards the sun before, is diminished, it will ascend therefrom like a stream, and carry along with it the reflecting particles of which the tail of the comet is composed; the impulse of the sun's light, as we have said, promoting the ascent.

But that the tails of comets do arise from their heads (p. 488), and tend towards the parts opposite to the sun, is farther confirmed from the laws which the tails observe; for, lying in the planes of the comets orbits which pass through the sun, they constantly deviate from the opposition of the sun towards the parts which the comets heads in their progress along those orbits have left; and to a spectator placed in those planes they appear in the parts directly opposite to the sun; but as the spectator recedes from those planes, their deviation begins to appear, and daily becomes greater. And the deviation, *cæteris paribus*, appears less when the tail is more oblique to the orbit of the comet, as well as when the head of the comet approaches nearer to the sun .; especially if the angle of deviation is estimated near the head of the comet. Farther; the tails which have no deviation appear straight, but the tails which deviate are likewise bended into a certain curvature; and this curvature is greater when the deviation is greater, and is more sensible when the tail, *cæteris paribus*, is longer; for in the shorter tails the curvature is hardly to be perceived. And the angle of deviation is less near the comet's head, but greater towards the other end of the tail, and that because the lower side of the tail regards the parts from which the deviation is made, and which lie in a right line drawn out infinitely from the sun through the comet's head. And the tails that are longer and broader; and shine with a stronger light, appear more resplendent and more exactly defined on the convex than on the concave side. Upon which accounts it is plain that the phenomena of the tails of comets depend upon the motions of their heads, and by no means upon the places of the heavens in which their heads are seen; and that, therefore, the tails of the comets do not proceed from the refraction of the heavens, but from their own heads, which furnish the matter that forms the tail; for as in our air the smoke of a heated body ascends either perpendicularly, if the body is at rest, or obliquely if the body is moved obliquely, so in the heavens, where all the bodies gravitate towards the sun, smoke and vapour must (as we have already said) ascend from the sun, and either rise perpendicularly, if the smoking body is at rest, or obliquely, if the body, in the progress of its motion, is always leaving those places from which the upper or higher parts of the vapours had risen before. And that obliquity will be less where the vapour ascends with more velocity, to wit, near the smoking body, when that is near the sun; for there the force of the sun by which the vapour ascends is stronger. But because the obliquity is varied, the column of vapour will be incurvated; and because the vapour in the preceding side is something more recent, that is, has ascended something more lately from the body, it will therefore be something more dense on that side, and must on that account reflect more light, as well as be better defined; the vapour on the other side languishing by degrees, and vanishing out of sight.

But it is none of our present business to explain the causes of the appearances of nature. Let those things which we have last said be true or false, we have at least made out, in the preceding discourse, that the rays of light are directly propagated from the tails of comets in right lines through the heavens, in which those tails appear to the spectators wherever placed; and consequently the tails must ascend from the heads of the comets towards the parts opposite to the sun. And from this principle we

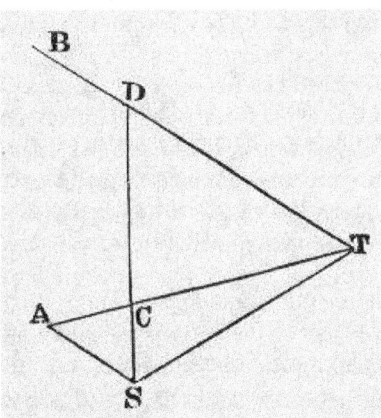

may determine anew the limits of their distances in manner following. Let S rep-resent the sun, T the earth, STA the elongation of a comet from the sun, and ATB the apparent length of its tail; and because the light is propagated from the extremity of the tail in the direction of the right, line TB, that extremity must lie somewhere in the line TB. Suppose it in D, and join DS cutting TA in C. Then, because the tail is always stretched out towards the parts nearly opposite to the sun, and therefore the sun, the head of the comet, and the extremity of the tail, lie in a right line, the comet's head will be found in C. Parallel to TB draw SA, meeting the line TA in A, arid the comet's head C must necessarily be found between T and A, because the extremity of the tail lies somewhere in the infinite line TB; and all the lines SI) which can possibly be drawn from the point S to the line TB must cut the line TA somewhere between T and A. Wherefore the distance of the comet from the earth cannot exceed the interval TA, nor its distance from the sun the interval SA beyond, or ST on this side the sun. For instance: the elongation of the comet of 1680 from the sun, Dec. 12, was 9°, and the length of its tail 35° at least. If, therefore, a triangle TSA is made, whose angle T is equal to the elongation 9°, and angle A equal to ATB, or to the length of the tail, viz., 35°, then SA will be to ST, that is, the limit of the greatest possible distance of the comet from the sun to the semi-diameter of the orbis magnus, as the sine of the angle T to the sine of the angle A, that is, as about 3 to 11. And therefore the comet at that time was less distant from the sun than by $^3/_{11}$ of the earth's distance from the sun, and consequently either was within the orb of Mercury, or between that orb and the earth. Again, Dec. 21, the elongation of the comet from the sun was $32^2/_3°$, and the length of its tail 70°. Wherefore as the sine of $32^2/_3°$ to the sine of 70°, that is, as 4 to 7, so was the limit of the comet's distance from the sun to the distance of the earth from the sun, and consequently the comet had not then got without the orb of Venus. Dec. 28, the elongation of the comet from the sun was 55°, and the length of its tail 56°; and therefore the limit of the comet's distance from the sun was not yet equal to the distance of the earth from the same, and consequently the comet had not then got without the earth's orbit. But from its parallax we find that its egress from the orbit happened about Jan. 5, as well as that it had descended far within the orbit of Mercury. Let us suppose it to have been in its perihelion Dec. the 8th, when it was in conjunction with the sun; and it will follow that in the journey from its perihelion to its exit out of the earth's orbit it had spent 28 days; and consequently that in the 26 or 27 days following, in which it ceased to be farther seen by the naked eye, it had scarcely doubled its distance from the sun; and by limiting the distances of other comets by the like arguments, we come at last to this conclusion, that all comets, during the time in which they are visible by us, are within the compass of a spherical space described about the sun as a centre, with a radius double, or at most triple, of the distance of the earth from the sun.

And hence it follows that the comets, during the whole time of their appearance unto us, being within the sphere of activity of the circum-solar force, and therefore agitated by the impulse of that force, will (by Cor. 1, Prop. XII, Book I, for the same reason as the planets) be made to move in conic sections that have one focus in the centre of the sun, and by radii drawn to the sun, to describe areas proportional to the times; for that force is propagated to an immense distance, and will govern the motions of bodies far beyond the orbit of Saturn.

There are three hypotheses about comets (p. 466); for some will have it that they are generated and perish as often as they appear and vanish; others, that they come from the regions of the fixed stars, and are seen by us in their passage through the system of our planets; and, lastly, others, that they are bodies perpetually revolving about the sun in very eccentric orbits. In the first case, the comets, according to their different velocities, will move in conic sections of all sorts; in the second, they will describe hyperbolas, and in either of the two will frequent indifferently all quarters of the heavens, as well those about the poles as those towards the ecliptic; in the third, their motions will be performed in ellipses very eccentric, and very nearly approaching to parabolas. But (if the law of the planets is observed) their orbits will not much decline from the plane of the ecliptic; and, so far as I could hitherto observe, the third case obtains; for the comets do, indeed, chiefly frequent the zodiac, and scarcely ever attain to a heliocentric latitude of 40. And that they move in orbits very nearly parabolical, I infer from their velocity; for the velocity with which a parabola is described is every where to the velocity with which a comet or planet may be revolved about the sun in a circle at the same distance in the subduplicate ratio of 2 to 1 (by Cor. VII, Prop. XVI); and, by my computation, the velocity of comets is found to be much about the same. I examined the thing by inferring nearly the velocities from the distances, and the distances both from the parallaxes and the phaenomena of the tails, and never found the errors of excess or defect in the veolocities greater than what might have arose from the errors in the distances collected after that manner. But I likewise made use of the reasoning that follows.

Supposing the radius of the orbis magnus to be divided into 1000 parts: let the numbers in the first column of the following table represent the distance of the vertex of the parabola from the sun's centre, expressed by those parts: and a comet in the times expressed in col. 2, will pass from its perihelion to the surface of the sphere which is described about the sun as a centre with the radius of the orbis magnus; and in the times expressed in col. 3, 4, and 5, it will double, triple, and quadruple, that its distance from the sun.

TABLE I.

The distance of a comet's perihelion from the Sun's centre.	The time of a comet's passage from its perihelion to a distance from the Sun equal to			
	The radius of the orbis magnus.	To its double.	To its triple.	To its Quadruple.
	d. h. ′	d. h. ′	d. h. ′	d. h. ′
0	27 11 12	77 16 28	142 17 14	219 17 30
5	27 16 07	77 23 14		
10	27 21 00	78 06 24		
20	28 06 40	78 20 13	144 03 19	221 08 54
40	29 01 32	79 23 34		
80	30 13 25	82 04 56		
160	33 05 29	86 10 26	153 16 08	?32 12 20
320	37 13 46	93 23 38		
640	37 09 49	105 01 28		
1280		106 C6 35	200 06 43	297 03 46
2560			147 22 31	300 06 03

[This table, here corrected, is made on the supposition that the earth's diurnal motion is just 59′, and the measure of one minute loosely 0,2909, in respect of the radius 1000. If those measures are taken true, the true numbers of the table will all come out less. But the difference, even when greatest, and to the quadruple of the earth's distance from the sun, amounts only to 16h. 55′.]

The time of a comet's ingress into the sphere of the orbis magnus, or of its egress from the same, may be inferred nearly from its parallax, but with more expedition by the following

TABLE II.

The apparent elongation of a comet from the sun	Its apparent diurnal motion in its own orbit.		Its distance from the earth in parts whereof the radius of the orbis magnus contains 1000.
	Direct.	Retrog	
60	2 18	00 20	1000
65	2 33	00 35	845
70	2 55	00 57	684
72	3 07	01 09	618
74	3 23	01 25	651
76	3 43	01 45	484
78	4 10	02 12	416
80	4 57	02 49	347
82	5 45	03 47	278
84	7 18	05 20	209
86	10 27	08 19	140
88	18 37	16 39	70
90	Infinite	Infinite	00

The ingress of a comet into the sphere of the orbis magnus, or its egress from the same, happens at the time of its elongation from the sun, expressed in col. 1, against its diurnal motion. So in the comet of 1681. Jan. 4, O.S. the apparent diurnal motion in its orbit was about 3° 5', and the corresponding elongation $71^2/_3°$; and the comet had acquired this elongation from the sun Jan. 4, about six in the evening. Again, in the year 1680, Nov. 11, the diurnal motion of the comet that then appeared was about $4^2/_3°$; and the corresponding elongation $79^2/_3$ happened Nov. 10, a little before midnight. Now at the times named these comets had arrived at an equal distance from the sun with the earth, and the earth was then almost in its perihelion. But the first table is fitted to the earth's mean distance from the sun assumed of 1000 parts; and this distance is greater by such an excess of space as the earth might describe by its annual motion in one day's time, or the comet by its motion in 16 hours. To reduce the comet to this mean distance of 1000 parts, we add those 16 hours to the former time, and subduct them from the latter; and thus the former be comes Jan. 4d. 10h. afternoon; the latter Nov. 10, about six in the morning. But from the tenor and progress of the diurnal motions it appears that both comets were in conjunction with the sun between Dec. 7 and Dec. 8; and from thence to Jan. 4d. 10h. afternoon on one side, and to Nov. 10d. 6h. of the morning on the other, there are about 28 days. And so many days (by Table 1) the motions in parabolic trajectories do require.

But though we have hitherto considered those comets as two, yet, from the coincidence of their perihelions and agreement of their velocities, it is probable that in effect they were but one and the same; and if so, the orbit of this comet must have either been a parabola, or at least a conic section very little differing from a parabola, and at its vertex almost in contact with the surface of the sun. For (by Tab. 2) the distance of the comet from the earth, Nov. 10, was about 360 parts, and Jan. 4, about 630. From which distances, together with its longitudes and latitudes, we infer the distance of the places in which the comet was at those times to have been about 280: the half of which, viz., 140, is an ordinate to the comet's orbit, cutting off a portion of its axis nearly equal to the radius of the orbis magnus, that is, to 1000 parts. And, therefore, dividing the square of the ordinate 140 by 1000, the segment of the axis, we find the latus rectum 19, 16, or in a round number 20; the fourth part whereof, 5, is the distance of the vertex of the orbit from the sun's centre. But the time corresponding to the distance of 5 parts in Tab. 1 is 27d. 16h. 7'. In which time, if the comet moved in a parabolic orbit, it would have been carried from its perihelion to the surface of the sphere of the orbis magnus described with the radius 1000, and would have spent the double of that time, viz., 55d. 8$\frac{1}{4}$h. in the whole course of its motion within that sphere: and so in fact it did; for from Nov. 10d. 6h. of the morning, the time of the comet's ingress into the sphere of the orbis magnus, to Jan. 4d. 10h. afternoon, the time of its egress from the same, there are 55d. 16h. The small difference of 7$\frac{3}{4}$h. in this rude way of computing is to be neglected, and perhaps may arise from the comet's motion being some small matter slower, as it must have been if the true orbit in which it was carried was an ellipsis. The middle time between its ingress and egress was December 8d. 2h. of the morning; and therefore at this time the comet ought to have been in its perihelion. And accordingly that very day, just before sunrising, Dr. Halley (as we said) saw the tail short and broad, but very bright, rising perpendicularly from the horizon. From the position of the tail it is certain that the comet had then crossed over the ecliptic, and got into north latitude, and therefore had passed by its perihelion, which lay on the other side of the ecliptic, though it had not yet come into conjunction with the sun; and the comet [see more of this famous comet, p. 475 to 486] being at this time between its perihelion and its conjunction with the sun, must have been in its perihelion a few hours before; for in so near a distance from the sun it must have been carried with great velocity, and have apparently described almost half a degree every hour.

By like computations I find that the comet of 1618 entered the sphere of the orbis magnus December 7, towards sun-setting; but its conjunction with the sun was Nov. 9, or 10, about 28 days intervening, as in the preceding comet ; for from the size of the tail of this, in which it was equal to the preceding, it is probable that this comet likewise did come almost into a contact with the sun. Four comets were seen that year of which this was the last. The second, which made its first appearance October 31, in the neighbourhood of the rising sun, and was soon after hid under the sun's rays, I suspect to have been the same with the fourth, which emerged out of the sun's rays about Nov. 9. To these we may add the comet of 1607, which entered the sphere of the orbis magnus Sept. 14, O.S. and arrived at its perihelion distance from the sun about October 19, 35 days intervening. Its perihelion distance subtended an apparent angle at the earth of about 23 degrees, and was therefore of 390 parts. And to this number of parts about 34 days correspond in Tab. 1 . Farther; the comet of 1665 entered the sphere of the orbis magnus about March 17, and came to its perihelion about April 16, 30 days intervening. Its perihelion distance subtended an angle at the earth of about seven degrees, and therefore was of 122 parts: and corresponding to this number of parts, in Tab. 1, we find 30 days. Again; the comet of 1682 entered the sphere of the orbis magnus about Aug. 11, and arrived at its perihelion about Sep. 16, being then distant from the sun by about 350 parts, to which, in Tab. I, belong $33\frac{1}{2}$ days. Lastly; that memorable comet of Regiomontanus, which in 1472 was carried through the circum-polar parts of our northern hemisphere with such rapidity as to describe 40 degrees in one day, entered the sphere of the orbis magnus Jan 21, about the time that it was passing by the pole, and, hastening from them towards the sun, was hid under the sun's rays about the end of Feb., whence it is probable that 30 days, or a few more, were spent between its ingress into the sphere of the orbis magnus and its perihelion. Nor did this comet truly move with more velocity than other comets, but owed the greatness of its apparent velocity to its passing by the earth at a near distance.

It appears, then, that the velocity of comets (p. 471), so far as it can be determined by these rude ways of computing, is that very velocity with which parabolas, or ellipses near to parabolas, ought to be described; and therefore the distance between a comet and the sun being given, the velocity of the comet is nearly given. And hence arises this problem.

PROBLEM.

The relation betwixt the velocity of a comet and its distance from the sun's centre being given, the comet's trajectory is required.

If this problem was resolved, we should thence have a method of determining the trajectories of comets to the greatest accuracy: for if that relation be twice assumed, and from thence the trajectory be twice computed, and the error of each trajectory be found from observations, the assumption may be corrected by the Rule of False, and a third trajectory may thence be found that will exactly agree with the observations. And by determining the trajectories of comets after this method, we may come at last, to a more exact knowledge of the parts through which those bodies travel, of the velocities with which they are carried, what sort of trajectories they describe, and what are the true magnitudes and forms of their tails according to the various distances of their heads from the sun; whether, after certain intervals of time, the same comets do return again, and in what periods they complete their several revolutions. But the problem may be resolved by determining, first, the hourly motion of a comet to a given time from three or more observations, and then deriving the trajectory from this motion. And thus the invention of the trajectory, depending on one observation, and its hourly motion at the time of this observation, will either confirm or disprove itself; for the conclusion that is drawn from the motion only of an hour or two and a false hypothesis, will never agree with the motions of the comets from beginning to end. The method of the whole computation is this.

LEMMA I.

To cut two right lines OR, TP, given in, position, by a third right line RP, so as TRP may be a right angle; and, if another right line SP is drawn to any given point S, the solid contained under this line SP 5 and the square of the right line OR terminated at a given point O, may be of a given magnitude.

It is done by linear description thus. Let the given magnitude of the solid be M^2 x N: from any point r of the right line OR erect the

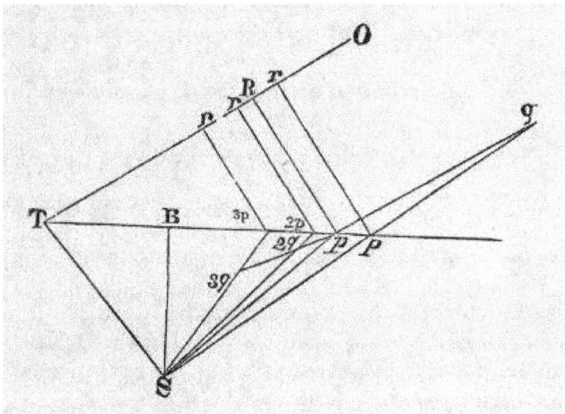

perpendicular rp meeting TP in p. Then through the point Sp draw the line Sq equal to M^2 x N
Or^2. In like manner draw three or more right lines S2q, S3q, &c.; and a regular line q2q3q, drawn through all the points q2q3q, &c., will cut the right line TP in the point P, from which the perpendicular PR is to be let fall. Q.E.F.

By trigonometry thus. Assuming the right line TP as found by the preceding method, the perpendiculars TR, SB, in the triangles TPR, TPS, will be thence given; and the side SP in the triangle SBP, as well as the error M^2 x N
Or^2 − SP. Let this error, suppose D, be to a new error, suppose E, as the error 2p2q ± 3p3q to the error 2p3p; or as the error 2p2q ± D to the error 2pP; and this new error added to or subducted from the length TP, will give the correct length TP ± E. The inspection of the figure will shew whether we are to add to or subtract; and if at any time there should be use for a farther correction, the operation may be repeated.

By arithmetic thus. Let us suppose the thing done, and let TP + e be the
correct length of the right line TP as found out by delineation: and thence
the correct lengths of the lines OR. BP, and SP, will be
OR − TR
TPe, BP + e, and $\sqrt{SP^2 + 2BPe + ee} = M^2N$
OR^2 + 2OR X TR
TPe + TR^2
TP^{2}ee.
Whence, by the method of converging series, we have
SP + BP
SPe + SP^2
2SP^{3}ee, &c., = M^{2}N
OR^2 + 2TR
TP x M^{2}N
OR^{3}e + 3TR^2
TP^2 x M^{2}N
OR^{4}ee
For the given coefficients M^{2}N
OR^2 − SP, 2TR
TP x M^{2}N
OR^3 − BP
SP, 3TR^2
TP^2 x M^{2}N
OR^4 − SB^2
2SP^3, putting F, F
G, F
GH, and carefully observing the signs, we find F + F
Ge + F
GHee = 0, and e + ee
H = − G. Whence, neglecting the very small term e^2
H, e comes out equal to −G. If the error e^2
H is not despicable, take

− G − G^2
H = e.

And it is to be observed that here a general method is hinted at for solving
the more intricate sort of problems, as well by trigonometry as by
arithmetic, without those perplexed computations and resolutions of
affected equations which hitherto have been in use.

LEMMA II.

To cut three right lines given in position by a fourth right line that shall pass through a point assigned in any of the three, and so as its intercepted parts shall be in a given ratio one to the other.

Let AB, AC, BC, be the right lines given in position, and suppose D to be the given point in the line AC. Parallel to AB draw DG meeting BC

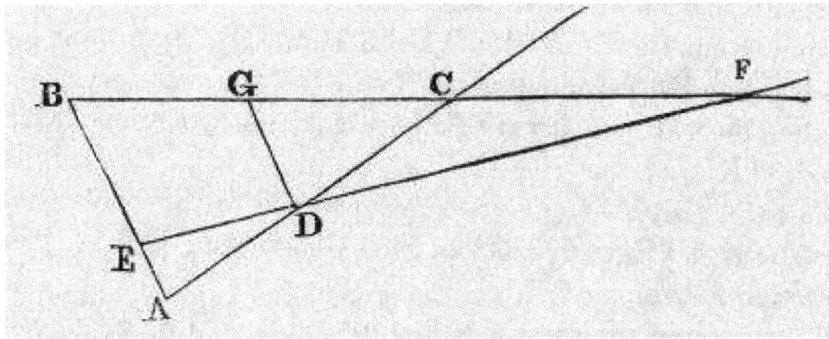

in G; and, taking GF to BG in the given ratio, draw FDE; and FD will be to DE as FG to BG. Q.E.F.

By trigonometry thus. In the triangle CGD all the angles and the side CD are given, and from thence its remaining sides are found; and from the given ratios the lines GF and BE are also given.

LEMMA III.

To find and represent by a linear description the hourly motion of a comet to any given time.

From observations of the best credit, let three longitudes of the comet be given, and, supposing ATR, RTB, to be their differences, let the hourly motion be required to the time of the middle observation TR. By Lem II. draw the right line ARB, so as its intercepted parts AR, RB, may be

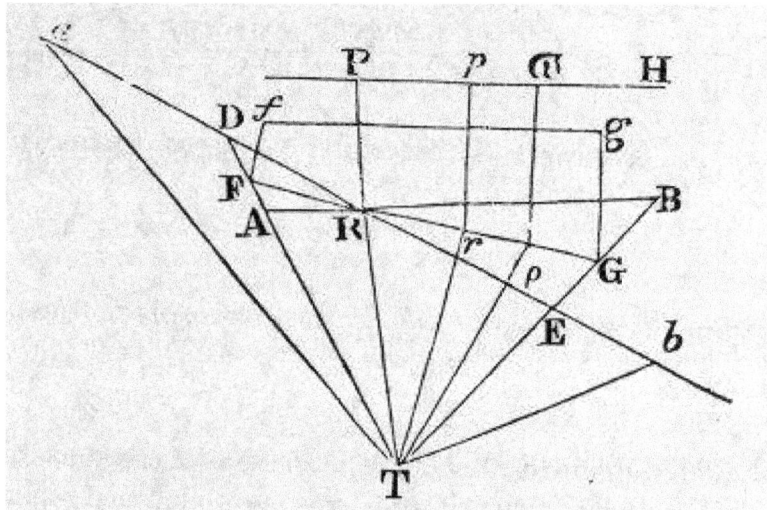

as the times between the observations; and if we suppose a body in the whole time to describe the whole line AB with an equal motion, and to be in the mean time viewed from the place T, the apparent motion of that body about the point R will be nearly the same with that of the comet at the time of the observation TR.

The same more accurately.

Let Ta, Tb, be two longitudes given at a greater distance on one side and on the other; and by Lem,. II draw the right line aRb so as its intercepted parts aR, Rb may be as the times between the observations aTR, RTA. Suppose this to cut the lines TA, TB, in D and E; and because the error of the inclination TRa increases nearly in the duplicate ratio of the time between the observations, draw FRG, so as either the angle DRF may be to the angle ARF, or the line DF to the line AF, in the duplicate ratio of the whole time between the observations aTB to the whole time between the observations ATB, and use the line thus found FG in place of the line AB found above.

It will be convenient that the angles ATR, RTB, aTA, BTb, be no less than of ten or fifteen degrees, the times corresponding no greater than of eight or twelve days, and the longitude taken when the comet moves with the greatest velocity for thus the errors of the observations will bear a less proportion to the differences of the longitudes.

LEMMA IV.

To find the longitudes of a comet to any given times.

It is done by taking in the line FG the distances Rr, Rρ, proportional to the times, and drawing the lines Tr, Tρ. The way of working by trigonometry is manifest.

LEMMA V.

To find the latitudes.

On TF, TR, TG, as radiuses, at right angles erect Ff, RP, Gg, tangents of the observed latitudes; and parallel to fg draw PH. The perpendiculars rp, ρω, meeting PH, will be the tangents of the sought latitudes to Tr and Tρ as radiuses.

PROBLEM I.

From the assumed ratio of the velocity to determine the trajectory of a comet.

Let S represent the sun; t, T, τ, three places of the earth in its orbit at equal distances; p, P, ω, as many corresponding places of the comet in

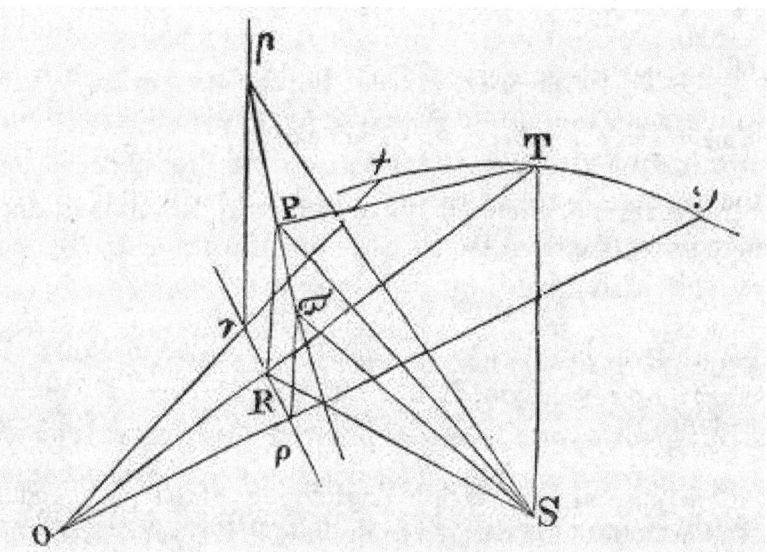

its trajectory, so as the distances interposed betwixt place and place may answer to the motion of one hour; pr, PR, ωρ, perpendiculars let fall on the plane of the ecliptic, and rRρ the vestige of the trajectory in this plane. Join Sp, SP, Sω, SR, ST, tr, TR, τρ, TP, and let tr, τρ, meet in O, TR will nearly converge to the same point O, or the error will be inconsiderable. By the premised lemmas the angles rOR, ROρ, are given, as well as the ratios pr to tr, PR to TR, and ωρ to τρ. The figure tTτO is likewise given both in magnitude and position, together with the distance ST, and the angles STR, PTR, STP. Let us assume the velocity of the comet in the place P to be to the velocity of a planet revolved about the sun in a circle, at the same distance SP, as V to 1; and we shall have a line pPω to be determined, of this condition, that the space pω, described by the comet in two hours, may be to the space V × tτ (that is. to the space which the earth describes in the same time multiplied by the number V) in the subduplicate ratio of ST, the distance of the earth from the sun, to SP, the distance of the comet from the sun; and that the space pP, described by the comet in the first hour, may be to the space Pω, described by the comet in the second hour, as the velocity in p to the velocity in P; that is, in the

72

subduplicate ratio of the distance SP to the distance Sp, or in the ratio of 2Sp to SP + Sp; for in this whole work I neglect small fractions that can produce no sensible error.

In the first place, then, as mathematicians, in the resolution of affected equations, are wont, for the first essay, to assume the root by conjecture, so, in this analytical operation, I judge of the sought distance TR as I best can by conjecture. Then, by Lem. II. I draw $r\rho$, first supposing R equal to $R\rho$, and again (after the ratio of SP to Sp is discovered) so as rR may be to $R\rho$ as 2SP to SP + Sp, and I find the ratios of the lines $p\omega$, $r\rho$, and OR, one to the other. Let M be to $V \times t\tau$ as OR to $p\omega$; and because the square of $p\omega$ is to the square of $V \times t\tau$ as ST to SP, we shall have, ex aequo, OR^2 to M^2 as ST to SP, and therefore the solid $OR^2 \times SP$ equal to the given solid $M^2 \times ST$; whence (supposing the triangles STP, PTR, to be now placed in the same plane) TR, TP, SP, PR, will be given, by Lem. I. All this I do, first by delineation in a rude and hasty way; then by a new delineation with greater care; and, lastly, by an arithmetical computation. Then I proceed to determine the position of the lines $r\rho$, $p\omega$, with the greatest accuracy, together with the nodes and inclination of the plane $Sp\omega$ to the plane of the ecliptic; and in that plane $Sp\omega$ I describe the trajectory in which a body let go from the place P in the direction of the given right line $p\omega$ would be carried with a velocity that is to the velocity of the earth as $p\omega$ to $V \times t\tau$. Q.E.F.

PROBLEM II.

To correct the assumed ratio of the velocity and the trajectory thence found.

Take an observation of the comet about the end of its appearance, or any other observation at a very great distance from the observations used before, and find the intersection of a right line drawn to the comet, in that observation with the plane Spω, as well as the comet's place in its trajectory to the time of the observation. If that intersection happens in this place, it is a proof that the trajectory was rightly determined; if otherwise, a new number V is to be assumed, and a new trajectory to be found; and then the place of the comet in this trajectory to the time of that probatory observation, and the intersection of a right line drawn to the comet with the plane of the trajectory, are to be determined as before; and by comparing the variation of the error with the variation of the other quantities, we may conclude, by the Rule of Three, how far those other quantities ought to be varied or corrected, so as the error may become as small as possible. And by means of these corrections we may have the trajectory exactly, providing the observations upon which the computation was founded were exact, and that we did not err much in the assumption of the quantity V: for if we did, the operation is to be repeated till the trajectory is exactly enough determined. Q.E.F.